MAPPING OF THE MOON

ASTROPHYSICS AND
SPACE SCIENCE LIBRARY

A SERIES OF BOOKS ON THE RECENT DEVELOPMENTS

OF SPACE SCIENCE AND OF GENERAL GEOPHYSICS AND ASTROPHYSICS

PUBLISHED IN CONNECTION WITH THE JOURNAL

SPACE SCIENCE REVIEWS

VOLUME 50

MAPPING
OF THE MOON

Past and Present

ZDENĚK KOPAL
Professor of Astronomy, University of Manchester, England

and

ROBERT W. CARDER
Chief, Aerospace Charting Branch, U.S. Defense Mapping Agency,
Aerospace Center, St. Louis, Missouri, U.S.A.

SPRINGER-SCIENCE+BUSINESS MEDIA, B.V.

Library of Congress Catalog Card Number 73–88592

ISBN 978-94-010-2135-7   ISBN 978-94-010-2133-3 (eBook)
DOI 10.1007/978-94-010-2133-3

TABLE OF CONTENTS

PREFACE

The aim of the present book has been to provide an outline – the first of its kind – of the history of the human efforts to map the topography of the surface of our satellite, from the days of pre-telescopic astronomy up to the present.

These efforts commenced modestly at the time when the unaided eye was still the only tool at the disposal of men interested in the face of our satellite; and were continued since for more than three centuries by a small band of devoted friends of the Moon in several countries. Many of these were amateur astronomers, and almost all were amateur cartographers; though some highly skilled in their art. The reader interested in the history of lunar mapping between 1600 and 1960 will find its outline in the first chapter of this book; and can follow the way in which the leadership in the mapping of the Moon, the cradle of which stood in Italy, passed successively to France, Germany, and eventually to the United States.

All efforts described in this chapter were wholly superseded by subsequent developments since 1960, largely motivated by logistic needs of a grand effort which culminated with repeated manned landings on the Moon between 1969–1972 – a feat which will remain for ever one of the glories of our century. Its urgency was created by the success of the initial Russian lunar missions in 1959 – the second of which (Luna 3) unveiled for us the principal topographic features of the Moon's far side for the first time. In the United States a comprehensive lunar mapping program was initiated in 1959 by two Department of Defense (DOD) mapping agencies – the U.S. Air Force Aeronautical Chart and Information Center (ACIC) and the U.S. Army Map Service (AMS). This coordinated DOD program was accomplished in support of, and funded by, the National Aeronautics and Space Administration (NASA).

In the course of this program – the greatest in the history of our subject – the cause of lunar mapping enlisted for the first time the services of professional cartographers experienced in the production of terrestrial maps of highest quality; and the outcome of their efforts speedily relegated all previous work into obsolescence. The methods and results of this work have been described in Chapters 6–9 of this book; while in the preceding Chapters 2–5 we laid down the underlying principles of physical selenodesy, including the definition of lunar coordinates and the methods for a determination of three-dimensional coordinates of lunar surface features.

The first to photograph the Moon's far side, and the first to compile a map showing some of the far side features, were the Russians. Their first map was published in 1960, and is described in Chapter 10 along with other U.S.S.R. lunar maps that were released throughout the 1960s. The final Chapter 11 gives account of one of the best and most popular lunar maps of the space-age, published in 1969 by the National Geographic Society of the United States.

Most of the illustrations and other supporting data related to the maps produced by the U.S. Government agencies in the 1960s were secured from ACIC, AMS, or NASA; and the authors are greatly indebted for a permission to reproduce them in this book. Especial appreciation is extended to Messrs William Cannell and Lawrence Schimerman of ACIC for their material help in Chapters 3 and 7. In addition, our thanks are due to Dr Yu. N. Lipsky of the Sternberg State Astronomical Institute in Moscow for providing many of the U.S.S.R. lunar maps which are reproduced in Chapter 10, and to Mr David Cook who willingly placed at our disposal the illustrations pertaining to the National Geographic Society Lunar map.

Last but not least, the authors take pleasure in expressing their sincere appreciation to Miss Ellen B. Finlay (now Mrs R. Carling) at Manchester, and Mr Joseph Lomuto of St. Louis, for invaluable editorial assistance in preparation of this book; and to Mrs Ruth Beesley (St. Louis) and Miss Fiona Holt (Manchester) who typed most part of the MS for the press.

CHAPTER 1

HISTORY OF LUNAR MAPPING: 1600–1960

The mapping of the Moon, or of any other celestial body, can be defined as a three-dimensional description – not interpretation – of permanent features which are characteristic of its surface. For the Earth, this task has been embarked upon close to the dawn of human civilization. The geographical horizons of our distant ancestors – increasing with the extent of their land- and seafaring – were put in a proper setting by the realization (sometime in the 5–4th centuries B.C.) that the Earth was a sphere, of dimensions which were (at least approximately) ascertained by Eratosthenes some time in the third century before our era.

In contrast with the Earth, the mapping of the Moon constitutes a subject the growth of which commenced at a much later date. The distance and dimensions of the Moon have, to be sure, been ascertained to a truly realistic approximation already by Aristarchos and Hipparchos in the 3rd and 2nd centuries B.C.*; and the title alone of Plutarch's tract 'De facie in Orbis Lunae' is a sufficient testimony of the fact that, for the unaided human eye, the Moon ceased to be a single picture-point long before the advent of telescopic astronomy.

No tracings of any details of the actual face of the Moon have, unfortunately, come down to us from antiquity.** Occasional references to Renaissance drawings of the Moon have failed to produce contemporary sources worthy of that name. This is, in particular, true of the supposed contributions by Leonardo da Vinci (1452–1519) – a great experimenter and scientist in addition to his immortal contributions to the world of art. His famous *Diaries* contain several references to his drawings of the Moon; but the latter have apparently all been lost.

In the course of the 15th and 16th centuries, a variety of theories attempted to account for the visible appearance of our satellite. Perhaps the most popular explanation was along the line that the lunar surface was smooth and polished like a mirror, and thus reflected an image of the Earth. Leonardo recognized clearly the fallacy of such an explanation; for in his *Diaries* he recorded that ... 'The Earth, when not covered by the water, presents different shapes from different vantage points; thus when the Moon is in the East, it would reflect other spots than when it stands overhead of in the West; whereas the spots upon the Moon, as seen at fullmoon, never change'.

* For an account of the methods of which they availed themselves for this purpose cf. Kopal, *The Moon* (sec. ed.), D. Reidel Publ. Co., Dordrecht, 1969; Chapter 5.
** A reference by Von Humboldt (1858) to a drawing of the Moon by Anaxagoras appears to be based on a misunderstanding of Plutarch's text of the *Life of Nicias*.

Moreover, 'a second reason is that an object reflected in a convex surface fills only a small part of the mirror, as is proved by perspective. The third reason is that when the Moon is full it only faces half the orb of the illuminated Earth'. This last reason of Leonardo's is, of course, fallacious (since the Moon at 'full' phase faces a 'new' Earth); but the former two are sufficient to dispose of the notion of a mirror-like Moon, which seems to have been so popular during the Renaissance and later that Galileo Galilei (see below) considered it necessary to refuse it again in 1610.

The first man of science – and one fully entitled to this epithet – who was sufficiently attracted by the appearance of the visible fact of the Moon to attempt a graphical representation of what he saw (and whose drawing was preserved for posterity) was William Gilbert (1540–1603), physician in-ordinary to Queen Elizabeth I of England and – his more important title to fame – the discoverer of terrestrial magnetism. In a book entitled *De Mundo Nostro Sublunari Philosophia Nova*, which remained unfinished at the time of his death and did not appear till in 1651 in Amsterdam through the good offices of James Boswell, Gilbert published what appears to be the first extant map of the Moon (see Figure 1.1), based on observations made with the naked eye and bearing some resemblance to the face of the Moon as we know it today. Since Gilbert died in 1603, this map must have been finished some time before that year – and, therefore, at least six years before the first recorded use of the telescope for astronomical purposes.

It is interesting to correlate the nomenclature Gilbert inscribed on his map with the terminology in modern usage. His 'Regio Magna Orientalis' (cf. Figure 1.1) coincides pretty well in position with our Mare Imbrium; while 'Regio Magna Occidentalis' represents a conglomerate of Mare Serenitatis, Tranquillitatis and Foecunditatis; and 'Britannia' appears to be Mare Crisium. On the other hand, 'Continens Meridionalis' and 'Insula Longa' are parts of our Oceanus Procellarum. It should be stressed that – in contrast to Plutarch and (later) Galileo, who considered the dark spots on the Moon to be seas – Gilbert (like his Italian contemporary, Giordano Bruno) regarded dark spots as continental ground, and bright spots as seas ('ita tellus erga Lunam maculas representat, terrarum continentium minus relucentium; aquarum vero et Oceani, propter laeviorem et luminis apprehensivam naturam magis splendentem'; cf. Gilbert, 1651; Chapter XIV, p. 173).

But the reasons which led Gilbert to construct his map of the Moon were not a mere curiosity. In another passage of his book he expressed a regret that no one did the same in antiquity, to enable him to find out whether after the lapse of almost twenty centuries the face of the Moon underwent any change – a surprisingly modern point of view! It would, perhaps, have consoled Gilbert if he knew that no such changes have been noted since the advent of the telescope up to the present time.

Although Gilbert's map of the Moon did not see the light of day till 1651, it must have been completed before 1603 – the year of his death – and as such it happens to be the first as well as the last contribution made by human hand to selenography in the days of pre-telescopic astronomy. Its existence remained generally overlooked; for at the time of its appearance it was no more than of historical interest. The book itself in which it appeared is now excessively rare – only two copies of it are known to exist in Great Britain: one at the British Museum, the other at the University Library in Glasgow; and it is from the latter that the map shown on Figure 1.1 has been reproduced.

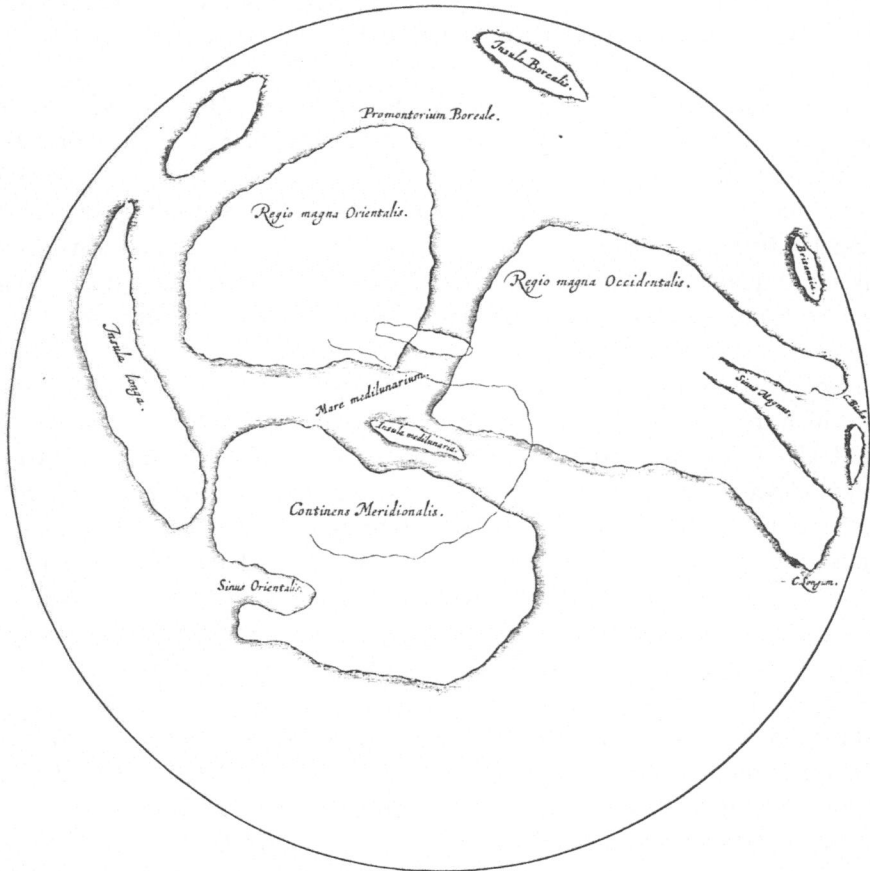

Fig. 1.1.   A drawing of the face of the Moon as it appeared to the naked eye, prepared by William Gilbert some time before his death in 1603, and facing p. 137 of his rare tract *De Mundo Nostro Sublunari*, etc. published (posthumously) in 1651 in Amsterdam. The present illustration is reproduced from a copy of this book in possession of the Glasgow University Library.

The advent of telescopic astronomy introduced also a new era in lunar mapping, inaugurated by Galileo Galilei (1564–1642) in 1609. Although Galileo was not the real inventor of the telescope, he was indubitably the first to use it for observations of celestial bodies; and as he recorded a year later in his *Nuncius Sidereus*, '... Sed missis terrenis ad coelestium speculationes me contuli: ac Lunam prius tam expropinquo sum intuitus, ac si vix per duas Telluria diametros abesset'*; and further (on p. 13) he continued that '... De facie autem Lunae quae ad aspectum nostrum vergit

---

* 'When I gave up observation of the terrestrial objects, I turned my attention to the celestial bodies: first I saw the Moon from such proximity as if it were barely two terrestrial diameters distant.'

primo loco dicamus; quam facillionis intelligentiae gratias in duas partes distinguo alteram nempe clariorem obscuriorem alteram... ut certo intelligamus, Lunae super-ficiem non perpolitam acquabilem exactissimaeque sphaericitatis existere ut magna Philosophorum cohors de ipsa deque reliquis corporibus coelestibus opinata est, sed contra inaequalem, asperam, cavitatibus tumoribus confertam, non secus ac ipsiusmet Telluris faciem, quae montium ignis vallumque profunditatibus hic inde distingui-tur'.*

As regards the 'spots' seen by him on the Moon, Galileo distinguished in his text between the 'old spots' (maculae antiquae) – i.e., the larger maria, visible to the naked eye and known to the ancient observers – and the 'new spots' (maculae novae) un-veiled by the telescope. Since his best *cannocchiale* (Galileo did not as yet call it a telescope) of 1610 was capable of resolving on the Moon details no smaller than about 30 km in size, Galileo's 'maculae novae' could have been large craters (like Clavius) and walled plaius (like Ptolamaeus, or Plato) which would have been visible through his optics.

Galileo's *Nuncius Sidereus* (1610) contained altogether five drawings of the Moon, two of which are reproduced on Figure 1.1. A mere glance at it will, however, convince us that Galileo was not a great astronomical observer; or else that the excitement of so many telescopic discoveries made by him at that time had temporarily blurred his skill or critical sense; for none of the features recorded on this (and other) drawings of the Moon can be safely identified with any known markings of the lunar land-scape.** In spite of the obvious shortcomings of these first telescopic observations of the Moon, their impact on the contemporary scientific thought was, however, pro-found; and promptly inaugurated the era of specifically selenographic literature and mapping, which it will be our aim to survey in this chapter.

Galileo did, to be sure, more than leave us only a few rather imperfect drawings of the face of our satellite. His inquisitive mind speculated already about the ways in which the altitudes of lunar mountains could be determined from characteristics of their illumination by the Sun. His observations were, however, weaker than theory; and he overestimated the heights of the lunar mountains 2–3 times.† In fact, the geometrical basis of lunar topography was not properly laid down till by Johannes Kepler (cf. his Letter to P. Guldin, published posthumously in 1634) some years later.

About simultaneously with Galileo's observations in Italy, the Moon came under telescopic scrutiny in England through the work of Thomas Harriot (1560–1621) – a

---

* 'Further, with regard to the side of the Moon facing us, let it be said first that one part of it is noticeably brighter, the other darker... so that we could perceive that the surface of the Moon is neither smooth nor uniform, nor very accurately spherical, as is assumed by a great many philosophers about the Moon and other celestial bodies; but that it is uneven, rough, replete with cavities and packed with protruding eminences, in no other wise than the Earth which is also characterized by mountains and valleys.'

** The ring-like configuration near the center of the apparent disc on Figure 1.2 (which Galileo compared with the central European land of Bohemia) may represent the crater Ptolemy.

† This notwithstanding the efforts of all his apologists in the years to come. Thus John Wilkins, in his tract on the *Discovery of a New World... in the Moon*, published (anonymously) in London in 1638; claimed (in propositio ix) that mountains as high as 9000 m exist on the Moon. In actual fact, the highest lunar mountains scarcely exceed the altitude of 5000 m.

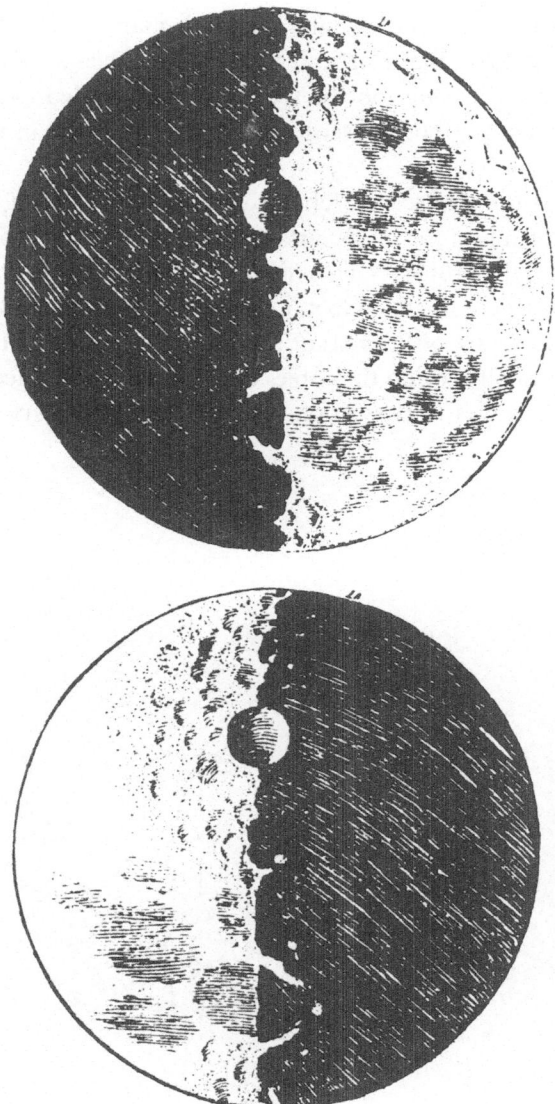

Fig. 1.2.   Two of the four drawings of the Moon made by Galileo Galilei and reproduced in his *Sidereus Nuncius* (Venice, 1610; also on p. 67 of vol. 3, 1 of his Edizione Nazionale, 1964). The drawings reproduce apparently but schematized views of what Galileo saw with his telescope; for none of the features recorded on them can be identified with certainty with any known formation.

rather shadowy figure among the British astronomers of the Elizabethan era. The results of his observations remained unpublished much longer than was the case with Gilbert. Their existence was first publicly noted by von Zach (1788) and Rigaud (1833); though it was not till eight years ago that a few of Harriot's drawings were at last reproduced by Strout (1965) from the MSS in possession of the Earl of Egremont; and the accompanying Figures 1.3 and 1.4 came from the same source.

A cursory comparison of Gilbert's and Harriot's maps a reproduced on Figures 1.1 and 1.4 makes it evident that Harriot's map required telescopic observations; for many details shown on it could not be seen with the unaided eye; therefore, it is certain that by September 1609 telescopes were already in use in Great Britain (cf. Mee, 1908).

Several recent historians of science (cf. e.g. Houzeau, 1882; Wolf, 1890) referred to the existence of drawings of the lunar surface in the volume *De Phoenomenis in Orbe Lunae*, etc., by La Galla (1612) which as regards publication date would be second in age to those of Galileo. In actual fact, none of the copies of La Galla's book preserved in the Royal Library of Brussels or the National Library of Florence contains any lunar drawings at all; nor do the copies extant in the Bibliothèque Nationale de Paris. However, the copies in possession of the National Library of Rome and of

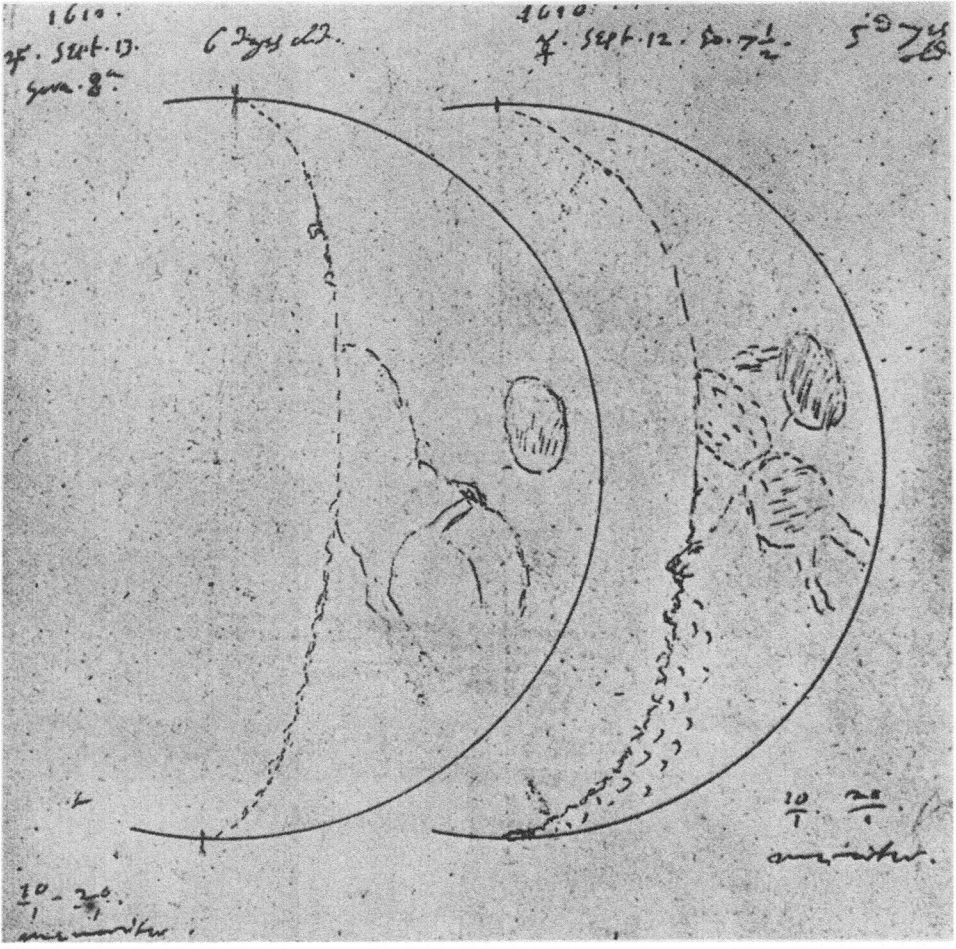

Fig. 1.3.   Two of the drawings of the waxing Moon by Thomas Harriot, dated 1610 September 12 and 13, based on telescopic observations, and showing clearly Mare Crisium as well as what appears to be outlines of Mare Tranquillitatis and Nectaris.

the Vatican Library does contain drawings of the Moon attributed previously to La Galla – but these are identical with those published previously by Galileo in his *Sidereus Nuncius*! This becomes intelligible when we realize that the publisher of La Galla's work was the same Tommasso Baglioni of Venice who published also Galileo's *Nuncius* only two years before. It is, therefore, probable that lunar drawings (obviously

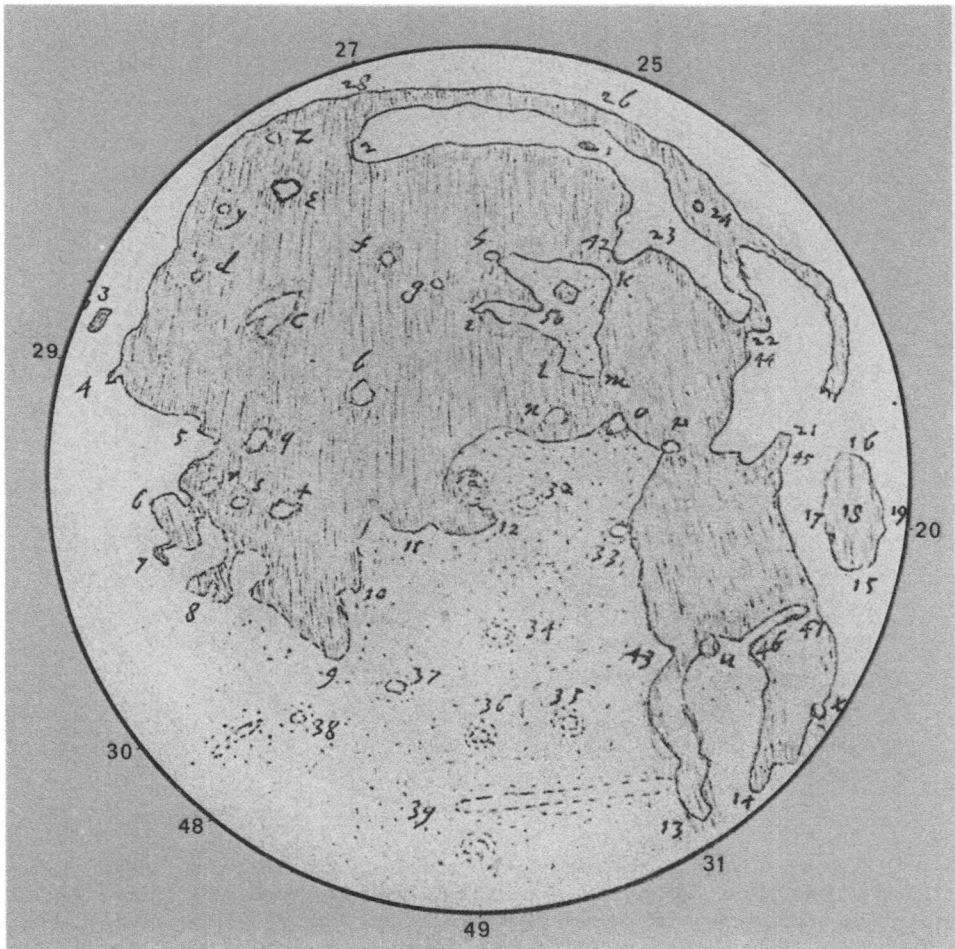

Fig. 1.4.   The first telescopic map of full Moon prepared by Thomas Harriot in England. Although, according to Rigaud (1833), Harriot commenced his observations of the Moon as early as July 1609, the present drawing probably followed the publication of Galileo's *Nuncius*. It is, however, superior in detail to all his drawings (Galileo did not draw full Moon) and contains many features which can be clearly recognized today. Note (under No. 39) a feature which appears to be one of the rays of Tycho.

Galileo's) were inserted in at least a part of the edition of La Galla's book (otherwise lacking any illustrations altogether) by the publisher himself – perhaps in order to increase its attractiveness for the more general reader.

Almost simultaneously with the contemporary work of Galileo and Harriot, telescopic drawings of the Moon were prepared in Germany by P. Christoph Scheiner, S.J. (1575–1650) of Ingolstadt – the historic adversary of Galileo Galilei in their subsequent dispute over sunspots – and published on p. 58 of his *Disquisitiones Mathematicae de Controversiis et Novitatibus Astronomicis* (Ingolstadt, 1614). From a technical point of view, Scheiner's map – reproduced on Figure 1.5 – goes a little beyond

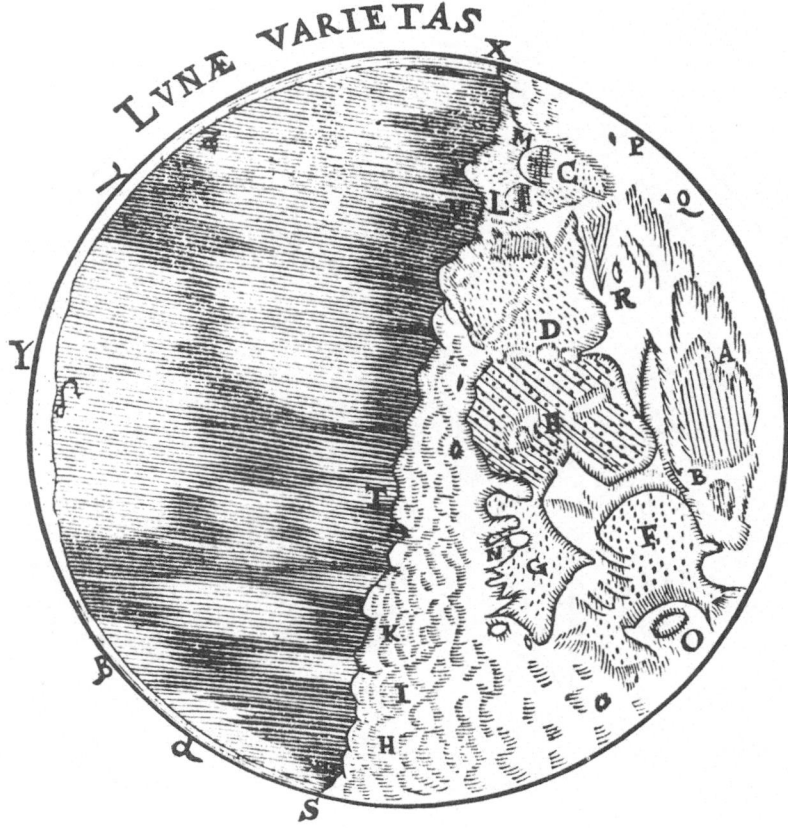

Fig. 1.5.   A drawing of the first quarter of the Moon, prepared by Christoph Scheiner probably before 1613, and published a year later in *Disquisitiones Mathematicae*. Some of the surface features labelled by letters can be readily identified. Thus A denotes Mare Crisium; E, Mare Tranquillitatis; F, Mare Foecunditatis; G, Mare Nectaris; and M, the crater Aristoteles. Like on all drawings reproduced on Figures 1–4, North is on top (as the Moon is seen by the naked eye, or the Galilean telescope). As Scheiner is known to have used the Keplerian (inverting) telescopes since 1613, it is believed that the drawing reproduced above was made before that time.

that of Harriot. Among subsequent early contributors to lunar topography the name of another Jesuit, the Belgian mathematician C. Malapert (1619) should now be mentioned; but like the essays of Galileo or Scheiner, his efforts remained limited to the depiction of a single phase on the scale of 4.9 cm (for a reproduction of this drawing cf., van de Vyver, 1971b; Figure 2).

In the ensuing years, sporadic attempts were made to construct maps of the Moon which should facilitate the determinations of terrestrial longitudes by simultaneous observations of lunar eclipses from different stations on the Earth. With this purpose in mind, the French astronomer Pierre Gassendi (1592–1655) and his friend Nicolas Claude Fabri de Peiresc (1580–1637) commenced lunar observations around 1618, and by 1634 were ready on their basis to construct a map. In order to do so, they enlisted the help of the engraver Claude Mellan (of whom we are going to hear more later on); and the results of this collaborative effort were published under the title *Phasium Lunae Icones* (Aix-en-Provence, 1636). An example of their drawings is reproduced on the accompanying Figure 1.6.

A whole cluster of new maps of the Moon appeared between 1644 and 1651. Thus A. Argoli, in his *Pandosium Sphaericum* (Padua, 1644) included a map of the lunar surface whose real author is, however, probably Francesco Fontana (1585–1656); for

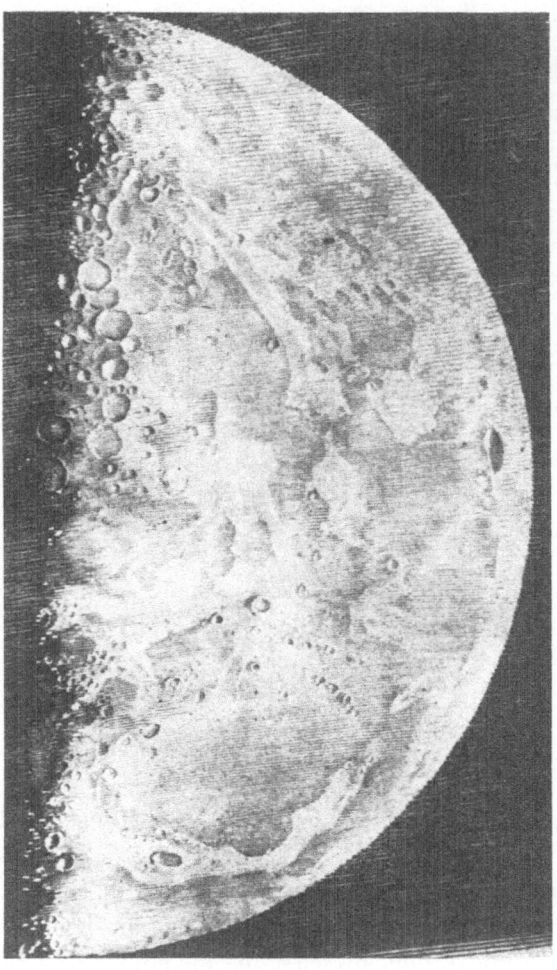

Fig. 1.6. An engraving of the last quarter of the Moon by Claude Mellan – one of three carried out around 1636.

– as was recently pointed out by Van de Vyver (1971a) – it differs only in some un-important details from one published by Fontana in his *Novae Coelestium Terrestrium-que Rerum Observationes* (Naples, 1646). Fontana's map (cf. Figure 1.7) must indeed be considered to be of considerably earlier origin than the date of its publication.

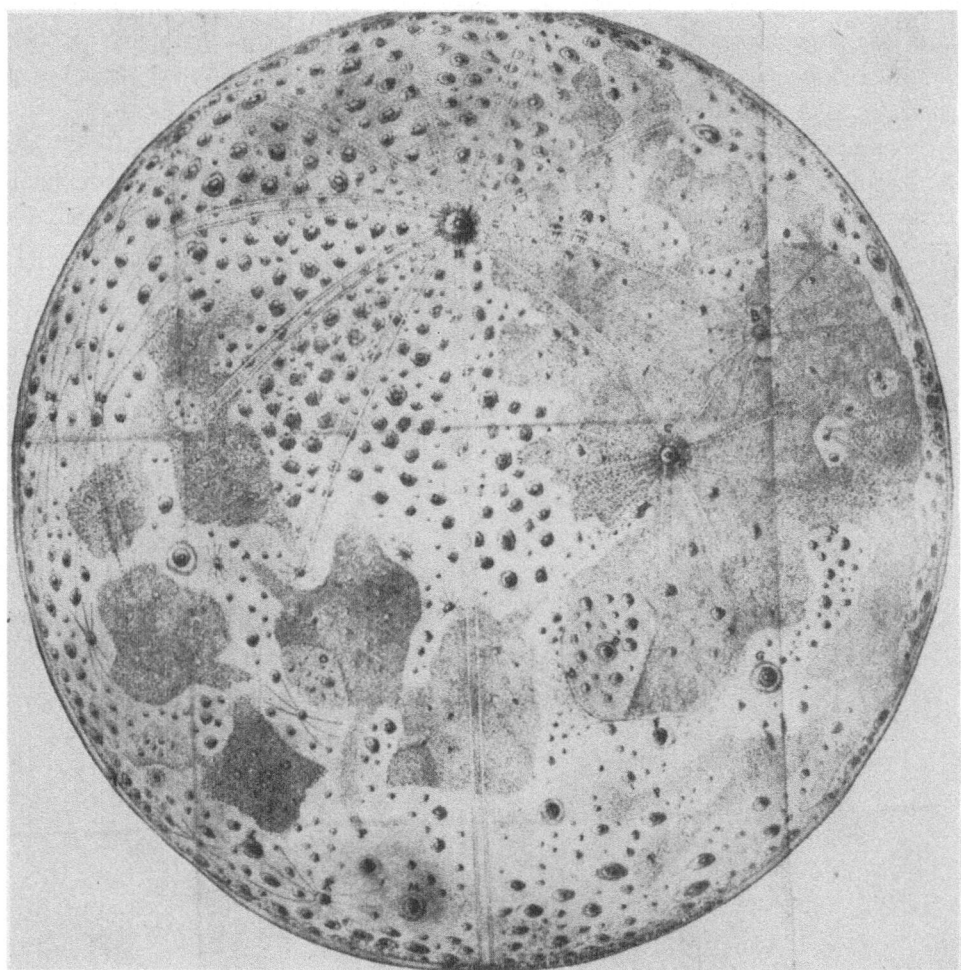

Fig. 1.7.   A map of the Moon by F. Fontana (1646).

Some drawings of the Moon included in his book go back to 1629 and were cir-culated prior to the appearance of his *Observationes* (as is attested by a letter to Galileo Galilei written by P. Fulgenzio Micanzio in 1638). Fontana's map is not very reliable in detail; and gives impression that its author, probably impressed by a great number of craters visible through his telescope, did not take trouble to ascertain their positions with any precision, but rather represented them as scattered more or less at random.

Another map of the Moon of not much greater scientific significance appeared in 1645, and its author was the Czech capucin P. Antonín Šírek z Rejty (1597–1660), better known under his latinized name of Anton Maria Schyrlaeus de Rheita. A keen scholar, he invented a reverting eyepiece for the Keplerian telescope; and also conjectured – almost a century before Halley – that fixed stars possessed proper motions of their own. The map of the Moon (18 cm in diam), which he constructed on the basis of his own observations, appeared as a part ('de facie lunae, etc.') of his large partly astronomical and partly theological work, entitled *Oculus Enoch et Eliae, Opus Theologiae, Philosophiae, et Verbi Dei Praeconibus utile et incundum* published in Antwerp in 1645 and dedicated to no one lesser than Jesus Christ. Rheita's map (reproduced on Figure 1.8) is rather scanty in detail, and contains no nomenclature; but the contours of the maria as well as the ray systems of Copernicus or Tycho are quite well represented. It is of interest to note that it is the first map of the Moon orientated with the south at the top (as seen through an inverting astronomical telescope).

The two most important contributors to selenography prior to 1650 were, however, M. F. van Langren and J. Hevelius. Michel Florent van Langren (1600–1675), a mathematician and cosmographer to King Philip IV of Spain, was – like Gassendi – very interested in the possibility of determining geographic longitudes at sea from lunar observations, and in 1645 completed a comprehensive map for this purpose. Moreover, in order to facilitate identification of individual details on the Moon, Langren was the first to describe the lunar face in terms of a comprehensive *nomenclature*, without which his map would indeed have been of very limited use for the intended purpose. A rudimentary nomenclature was, to be sure, introduced to the Moon already by Gilbert; and alphabetic letters (or numbers) were used, to a limited extent, by Harriot or Scheiner. However, Langren was the first to use personal names for this purpose. The names he attached – rather inauspiciously – to 322 formations known to him, were mostly those of the Spanish and Austrian kings, princes and churchmen (the crater now well known as Copernicus was given by Langren the name of his royal patron Philip IV, and the present Oceanus Procellarum was called Oceanus Phillipicus; while Mare Imbrium for him was Mare Austriacum) – a policy of ingratiation to be expected from a courtier.

Moreover, to ensure a priority for his nomenclature, Langren obtained for his map a Royal Privilege (dated 3 March 1645 at Brussels, then in possession of Spanish Crown), under which 'it was prohibited by Royal Diploma to change the names of this map under pain of indignation, and also to make in any way copies under pain of confiscation and a fine of three florins'. Nevertheless, in spite of the Royal Diploma and its stipulations, very little of Langren's nomenclature survived to this day (on the same place) – mainly the name Sinus Medii in central part of the apparent disc of the Moon, and that of a large peripheral crater Langrenus which its author named after himself.

Langren's map appeared in Brussels in 1645 under the name of 'Lumina Austriaca Philippica'. Only four copies of this very rare map (located at Paris, Leiden, Edinburgh

Fig. 1.8.   A map of the Moon by A. M. de Rheita (1645).

and San Fernando) survived up to the present time. The specimen (34.4 cm in diam) from the Bibliothèque Nationale de Paris is reproduced on Figure 1.9.

Johannes Hevelius (1611–1687), Burgomaster of Danzig, entered the history of lunar mapping through the publication of his book *Selenographia sive Lunae De-*

Fig. 1.9.   A map of the Moon prepared by M. F. van Langren, as it appeared in his *Selenographia Langreniana* (1645). Reproduced from the Paris copy of the original.

scriptio (Danzig, 1647), which contains three general maps (of 28.5 cm in diam) and 40 drawings of individual phases of the Moon (each 16.3 cm in diam) accompanied by explanatory text. Of the maps two are illustrative (see Figure 1.10); but the third – more schematic – contains the nomenclature proposed by Hevelius for the principal formations on the lunar surface independently of Langren; and paying no attention

Fig. 1.10.   One of the three lunar maps prepared by Johannes Hevelius in 1645, as it appeared in his *Selenographia* (Danzig, 1647). It takes account for the first time of the phenomenon of libration.

to the Spanish King's Royal Diploma protecting his 'rights' on the Moon. This map has been reproduced on the accompanying Figure 1.11. Although very little of the nomenclature proposed by Hevelius survived likewise to this day, one cannot bypass without mention the marvellous example of serendipity which led Hevelius to bestow on the southern shores of the present Mare Tranquillitatis – the region where Apollo 11 landed on the Moon in July 1969 – the name 'Apollonia'.

Hevelius's maps reproduced on Figures 1.10 and 1.11 show (by their double limb contours) for the first time those parts of the lunar limb which become visible only during extreme librations of our satellite (of which Hevelius was the co-discoverer).

For over thirty years since its publication Hevelius's *Selenographia* remained un-surpassed in its field. But, unfortunately, no more prints of its maps were made during the lifetime of their author; and after his death, one of the copper engravings of the maps was reportedly made to serve as the bottom of a tea-pot.

In 1649 – shortly after the appearance of Hevelius's *Selenographia* – another map

Fig. 1.11. Another map of the Moon by Hevelius, more schematic than the first, but containing his nomenclature.

was published by Eustachio Divini (1610–1685) in Rome, and the reader will find its reproduction on Figure 1.12. Although Divini claimed to have used a micrometer in its preparation, a scrutiny of his map fails to show any apparent benefit of it; and the map itself possesses such resemblance to that of Hevelius that it must have been at least influenced by it. The same is even more obviously true of another map pub-lished by Pierre Borel (1610–1671) in his tract *De Vero Telescopii Inventore etc.* (Hague, 1655). As this map has generally excaped the attention of lunar historians, it is being reproduced on the accompanying Figure 1.13; but its comparison with

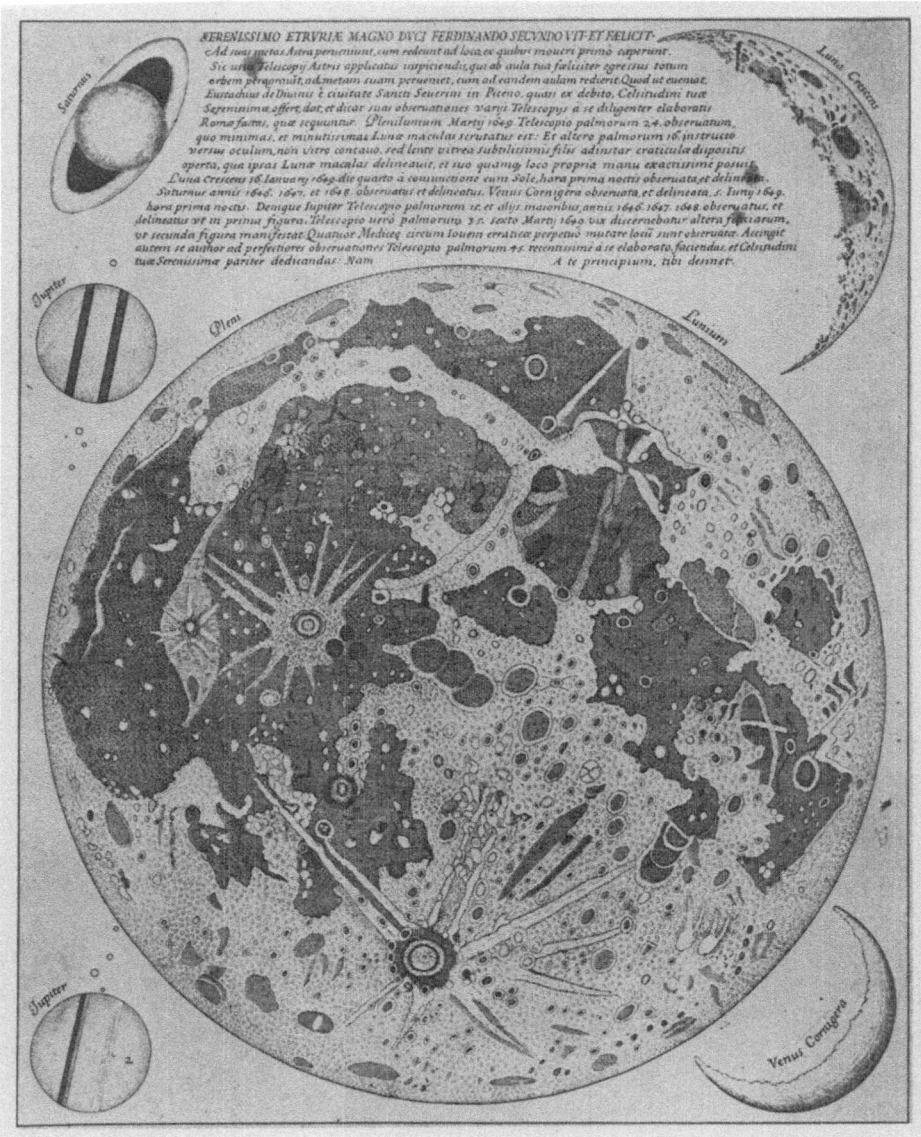

Fig. 1.12.   A map of the Moon by E. Divini (1649).

Hevelius's maps reproduced on Figure 1.10 and 1.11 speaks for itself (except that Borel reversed their orientation and placed North on top).

In 1651, a much more important map was published by Giovanni Battista Riccioli (1598–1671), S.J., from Ferrara, in his large work *Almagestum Novum* (Bologna, 1651), the chapter of which concerning the Moon contains two maps (28 cm in diam) – see Figures 1.14 and 1.15 – one of which (reproduced on Figure 1.15) shows the effects

experientia & memoria satis certa mihi dictavit, ute-
re si lubet. Dabantur Lutetiis nona die mensis Ju-
lii Anno 1655.

Johannes filius Zachariæ Joannidis primi Inventoris Te-
lescopii Middelburgensis sub tabula phaseos Lunæ, quam Telescopio suo
sæpe vidit, notat hæc quæ sequuntur.

EGo diversis temporibus Lunam inspexi Seleno-
scopio meo, cum plena esset, adeoque inveni
semper in infima parte disci Lunaris parvum quem-
dam globulum, sive sphærulam vergentem ferè ad
fundum sive infimam partem Lunæ, non in medio
disci, sed paululùm declinantem versus latus dex-
trum Lunæ, sicuti patet ex schemate quod exhibeo.
Sphærula illa ipsa etiam plena erat maculis sparsis
hinc inde lucidis, & immixta luminibus sicuti ipsum
corpus aut discus totus Lunæ. Sed in medio sphærulæ istius globuli aperiebat sese parvum & exi-
guum punctum ceu centrum præter modum & ex-
tensè lucidum : Ex quo sphærulæ puncto oriuntur
& procedunt sex lineæ lucidæ tamquam crenæ aut
sulci, sicuti in peponis viridis cortice visuntur, quæ
ex inferiori parte, videlicet ex sphærula illa sursum
& deorsum tendunt tanquam radii lumine eximio
inter maculas insignes, quas lineas vix videas aut de-
prehendas nisi cœlum admodum sit tersum & sudum
valdè. Dicere hic & annotare maculas in universo
disco lunari meum non est, cum iis vacare dignè non
potui, neque maculæ illæ eâdem formâ eædem semper maneant, nam operam meam solùm dedi ut Te-

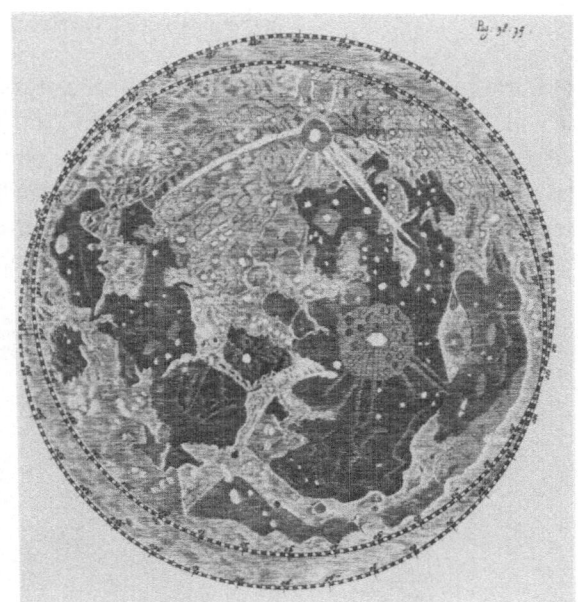

Fig. 1.13.   A map of the Moon by Pierre Borel (1655).

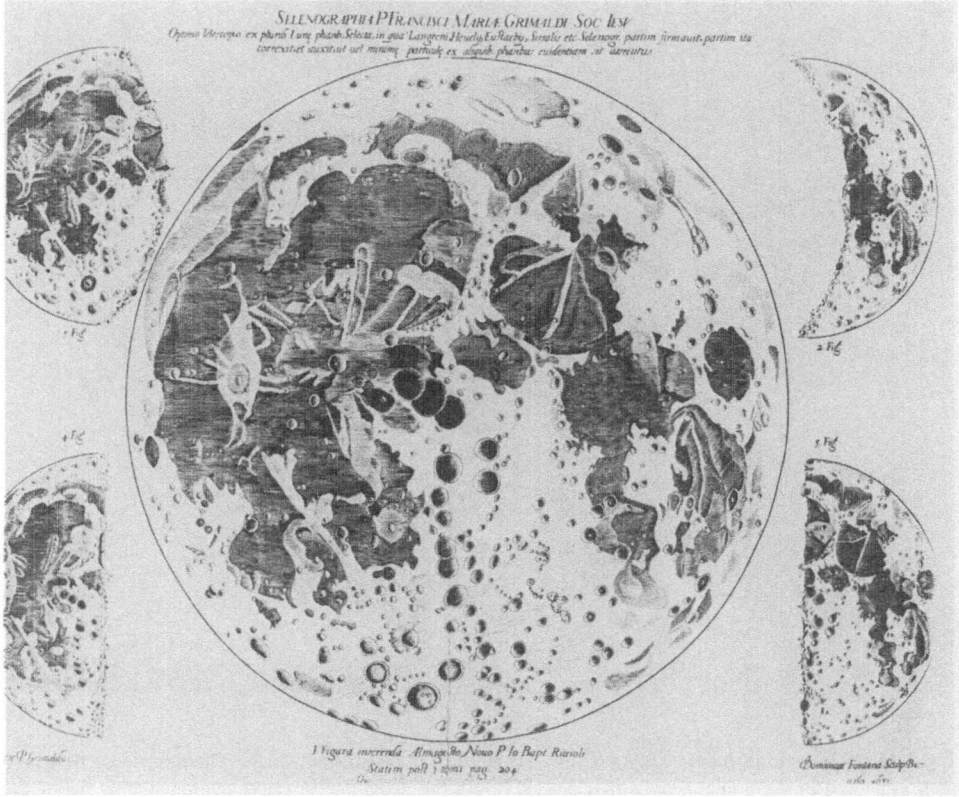

Fig. 1.14.   A map of the Moon reproduced opposite p. 204 of Riccioli's *Almagestum Novum* (Bologna, 1651), due to F. M. Grimaldi.

of librations and contains the lunar nomenclature, as introduced by Riccioli, much of which is still in use today. In particular, Riccioli (like Langren) referred to the lunar large dark patches as 'maria', and the smaller ones as 'palludes' – misnomers which have remained in use to this day – while the term 'terrae', given by him to the bright regions, did not take root in general usage. However, such familar by words as Mare Serenitatis, Oceanus Procellarum, and many others, we owe to Riccioli. He also fol-

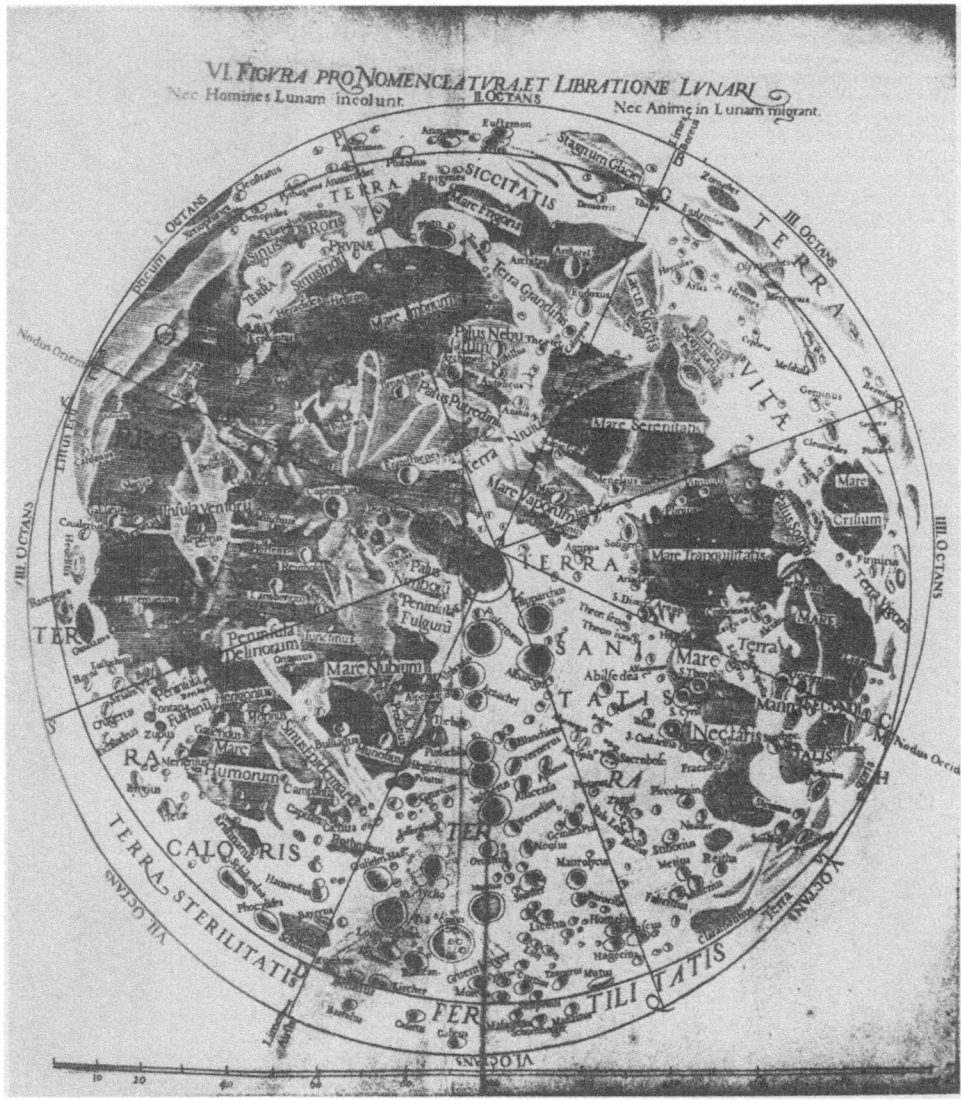

Fig. 1.15. The second lunar map by F. M. Grimaldi (taking account of the phenomena of libration) as it appeared in Riccioli's *Almagestum Novum*. Note the delightful inscription immediately below the heading: 'Nec Homines Lunam incolunt, nec Animae in Lunam migrant' – a prophesy which the events of the past few years have at least partly proved false.

lowed the custom of designating lunar craters by proper personal names, but wisely substituted those of the scholars for those of contemporary potentates; and thus divorced the face of the Moon from Langren's courtoisie. In particular, Riccioli used the names of the scholars of antiquity for craters located in the northern hemisphere of the Moon (after grouping the disciples around their respective masters), and placed the names of the Renaissance scholars to the south – a division which has not subsequently been too closely respected. The maps included in the *Almagestum* were not the work of Riccioli himself, but of his collaborators Francesco Maria Grimaldi (1618–1663) and C. Sirsalis.

Another contribution of some importance to lunar topography was made in 1660 by Geminiano Montanari (1633–1687) in Modena, discoverer of the light changes of Algol. Little known even at the time of its publication in Malvasia's *Ephemerides* (1662), its existence was virtually forgotten for more than three centuries – until its discovery and re-publication by Bonacini (1931). This map (38 cm in diam on the original) is reproduced in Figure 1.16, and was based on Montanari's observations with a telescope of 6.3 m focal distance carried out in October 1662. According to Malvasia, a micrometer was used in plotting the map.

If Montanari's map represented an original contribution to selenography, this was

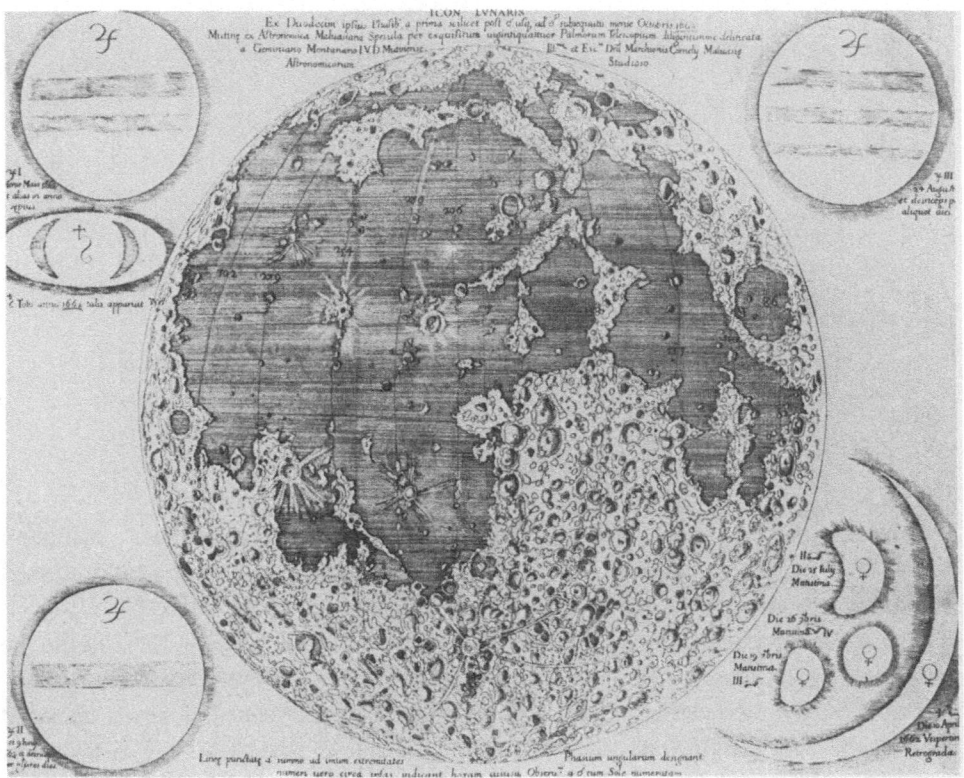

Fig. 1.16. A map of the Moon by G. Montanari (1662), discoverer of the light changes of Algol.

not true of the map that appeared in Athanasius Kircher's *Mundus Subterraneus* (Amsterdam, 1664). This work, (published also in Rome), contains a lunar map 29.9 cm in diam (see Figure 1.17), which bears so marked a resemblance to Rheita's map of 1645 that it may have been merely copied from it. Moreover, in the 1678

Fig. 1.17. A map of the Moon published by P. A. Kircher (1660).

Amsterdam edition of Kircher's book, the map is attributed to Scheiner – although there is nothing else to connect it with the learned Jesuit of that name who produced a rudimentary lunar map in 1614 (see Figure 1.5), and who died in 1650.

The contribution of another Jesuit, P. Valentin Stansel (1621–1705) should be mentioned in chronological order in this connection. A native of Olomouc, Czechoslovakia (where, two centuries later, J. F. J. Schmidt commenced work on his great map of the Moon published in 1878), he spent a major part of his life in Brazil as lecturer in theology at San Salvador. In his booklet *Propositiones Selenographicae, sive de Luna* (1655) Stansel inserted a drawing of full Moon, 7 cm in diam, apparently original but of no particular scientific importance – a drawing recently reproduced by van de Vyver (1971b). A somewhat more interesting map was produced at about the same time by the Capuchine monk Michel Lasséré (1613–1697), better known under his Order name of Cherubin d'Orléans. In his book *La Dioptrique Oculaire* (1671), Cherubin published two lunar maps, which were likewise recently reproduced by van de Vyver (1971b).

Towards the end of the 17th century, all previous selenographic efforts were overshadowed by the great map of Giovanni Domenico Cassini (1625–1712), prepared between 1671 and 1679 from Cassini's drawings supplemented by those of Leclerc and Patigny. Its reproduction as shown on Figure 1.18 reveals several well-known formations which appeared on it for the first time in the history of lunar mapping – such as the Altai mountains, Mare Smythii, or the Rheita valley. One conspicuous formation is, however, still missing: namely, the Alpine valley, drawn for the first time

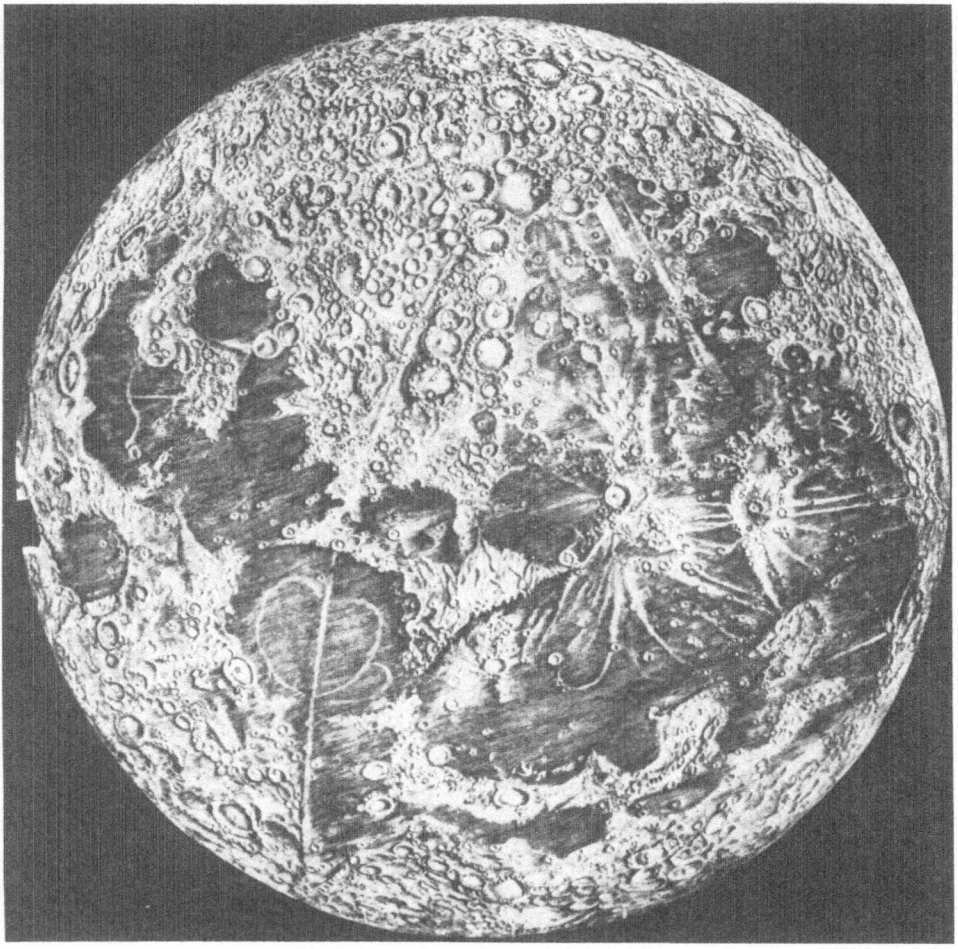

Fig. 1.18.   A map of the Moon by J. D. Cassini from 1680, engraved by C. Mellan.

by Francesco Bianchini (1662–1729) and published in his *Hespheri et Phosphori Nova Phaenomena* (1728).

Tradition has it that Cassini's map was engraved by the same Claude Mellan, who engraved his own drawings of the Moon with Gassendi around 1636 (cf. Figure 1.6). By 1680 Mellan was 82 years of age, though he still had 8 years to live. Of the actual engraver we have, however, no positive knowledge. An ancient inscription penciled on the verso side of the very rare print of this map in possession of the Royal Astronomical Society in London (one of the two in existence; the other being in the Observatoire National de Paris) merely mentions 'engraved by an Italian artist'. This could, however, well apply to Mellan, who spent many years in Italy, and imported the engraving techniques from there to France.

Only a few copies of the original edition of this map were apparently ever printed. In 1785 – more than 100 years later – the copper plate of the engraving was found at the Imprimerie Royale in Paris, and new reprint of it made. But in the turmoil of the revolutionary years no one thought of bringing the plate back to the Observatory; and a search for it made in 1828 revealed that it was sold by the Imprimerie in the meantime to a coppersmith – conceivably to serve the same profane purpose as did the copperplate with the engraving of Hevelius's map a century and a half before.

A glance at Cassini's map reproduced on Figure 1.18 shows it to be of a different class in accuracy from all its predecessors, but still rather deficient in impersonal representation of details required of mapping in more modern times. For it is obvious that its author (or Mellan) took more than one 'poetical license' in depicting (or engraving) the stony face of the Moon – such as reflected in the (non-existing) double loop of Tycho's bright ray passing through Mare Serenitatis; or the lovely outline of angel's head at the Heracleides Promontory of the southern shores of Sinus Iridum.

The achievements of Cassini and his contemporaries in the field of lunar topography become all the more impressive when we stop to consider the meager telescopic means at their disposal for the observations of our satellite. Throughout that period we are still in the first geological age of optical dinosaurs, characterized by small heads on huge bodies. The apertures of their simple objective lenses seldom exceeded 6–8 in.; but their focal ratios were excessively large (to lessen chromatic aberration), resulting in focal lengths unequally by almost any telescopes built since. Thus the instrument with which Hevelius carried out most of his observations of the lunar surface at Danzig possessed a focal length of 49 m (cf. Figure 1.19).

Needless to say, such telescopes could be but very crudely mounted. They had no tubes, and their objectives were mostly fixed at the end of a long pole, directed to different parts of the sky by the pull of ropes. Sometimes, in desperation, the astronomer dispensed with the mounting altogether, fixed his objective on to the roof of a building, and waited for a transit of his celestial object on the ground, with an eyepiece in his hand (see Figure 1.20). This was indeed the accepted practice of telescopic work at the Paris Observatory in the days of J. D. Cassini; and most part of the details on the lunar surface as shown on his map on Figure 1.18 were obtained in this way.

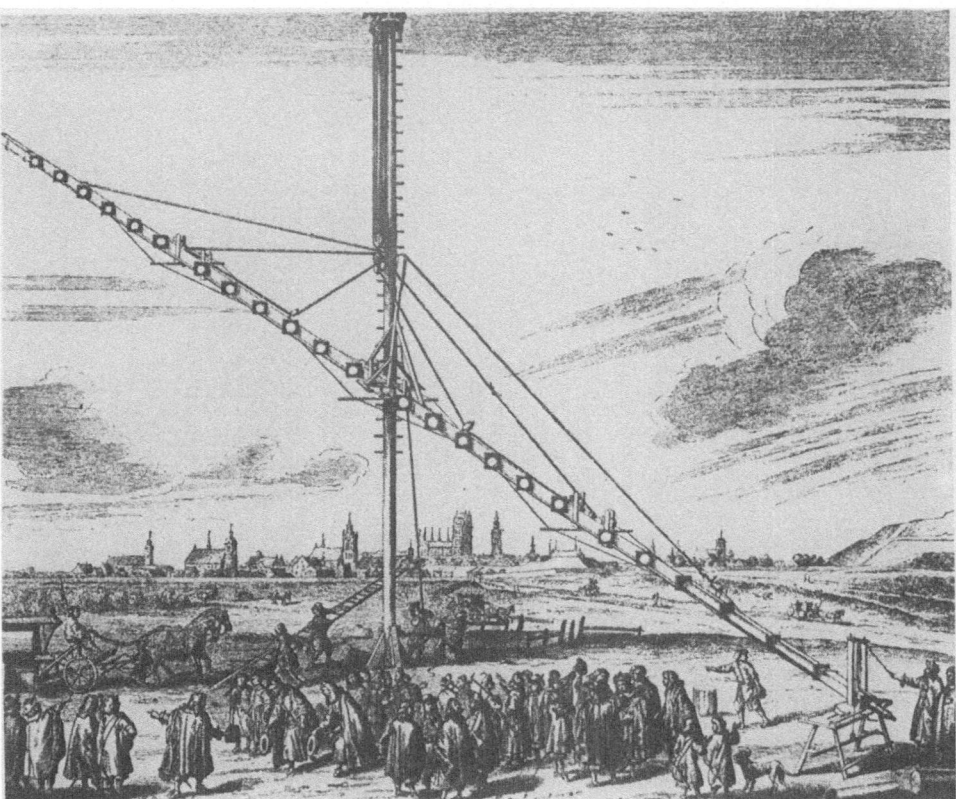

Fig. 1.19.   A telescope used by Johannes Hevelius in Danzig, used for his observations of the Moon in the middle of the 17th century. His maps reproduced on Figures 1.10 and 1.11 resulted from the observations made with such means.

After Cassini, no real further progress in selenography was made for many years. To be sure, in 1694 Georg Christoph Eimmart (1638–1705) engraved and published a 28-cm map of the Moon (for its recent reproduction see van de Vyver, 1971b) which is, however, relatively primitive in comparison with Cassini's; and the same remains to be true also of a map published in 1764 by the Jesuit Father Maximilian Hell (1720–1792), a Slovak-born director of the Vienna Observatory, known mainly for his observations of the transit of Venus across the Sun in 1768. His map – 18.5 cm in original – added little to our knowledge of the Moon; but on account of its scarcity we reproduce it on the accompanying Figure 1.21.

In the first half of the 18th century, the age of the long-necked telescopic Dinosaurs of the earlier epoch gradually came to an end (particularly, since Dollond's discovery of the achromatic objective in 1759); and this fact is also fully reflected in the selenographic literature of that time – or rather in the lack of it. For since about the turn of the 18th century the production of new lunar maps came temporarily to a standstill. In particular, on none of the maps which we had so far an opportunity to mention

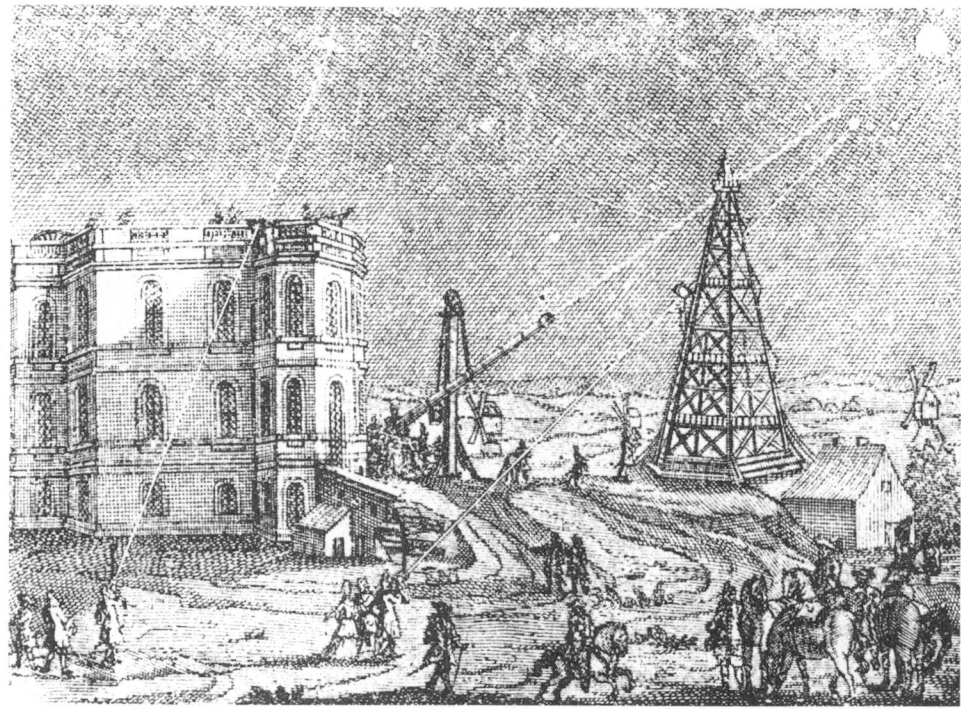

Fig. 1.20.   A night scene (engraved by a contemporary artist) of observing activity at the Paris Observatory in the days of J. D. Cassini. The data for his map of the Moon as reproduced on Figure 15.14 were obtained under such conditions.

did any system of lunar *coordinates* make its appearance – a concept fundamental to modern terrestrial map-making; but one which on the Moon took a long time to crystallize from contemporary efforts. The essential elements for the emergence of such a system – namely, the position of the equation and direction of the poles on the Moon – were established (at least approximately) by astronomers of the 17th century from the observed librations of the lunar globe (discovered by Gassendi and Hevelius before 1650), on which more will be said in the next chapter. At present we merely wish to stress that more than hundred years had to come and go since the discovery of lunar librations before the selenographers made appropriate use of them on their maps; and the first man to do so was Tobias Mayer (1723–1762).

Johann Tobias Mayer – the veritable father of lunar cartography as we understand this term today – was born in Marbach near Stuttgart in Germany on 17 February 1723, and spent the rest of his lamentably short life between the neighbourhood of his native region and Göttingen, where he died on 20 February 1762. As a young man he received some instruction in military cartography as well as engraving, both of which he put since 1748 to good use for the construction of the first real maps of our satellite. His motivation – like with Langren more than a century ago – was mainly geodetical; and the aim was to determine geographical longitudes from the times of entry of different craters in the Earth's shadow during total eclipses of the Moon.

The earlier maps available before Mayer were inadequate for this purpose. In order to lessen their deficiencies, Mayer embarked on the construction of his own maps, based on a micrometric triangulation of the Moon's surface. His first step was to investigate (in his *Bericht von den Mondskugeln*, Nürnberg, 1750) the rotation of the lunar globe from its observed librations, and found its axis of rotation to be inclined to the ecliptic by 1°29′ (the modern value of this angle being 1°31′). Moreover, for the fundamental control point of his system of coordinates Mayer adopted the crater Manilius near the center of the apparent disc of the Moon, to which he assigned

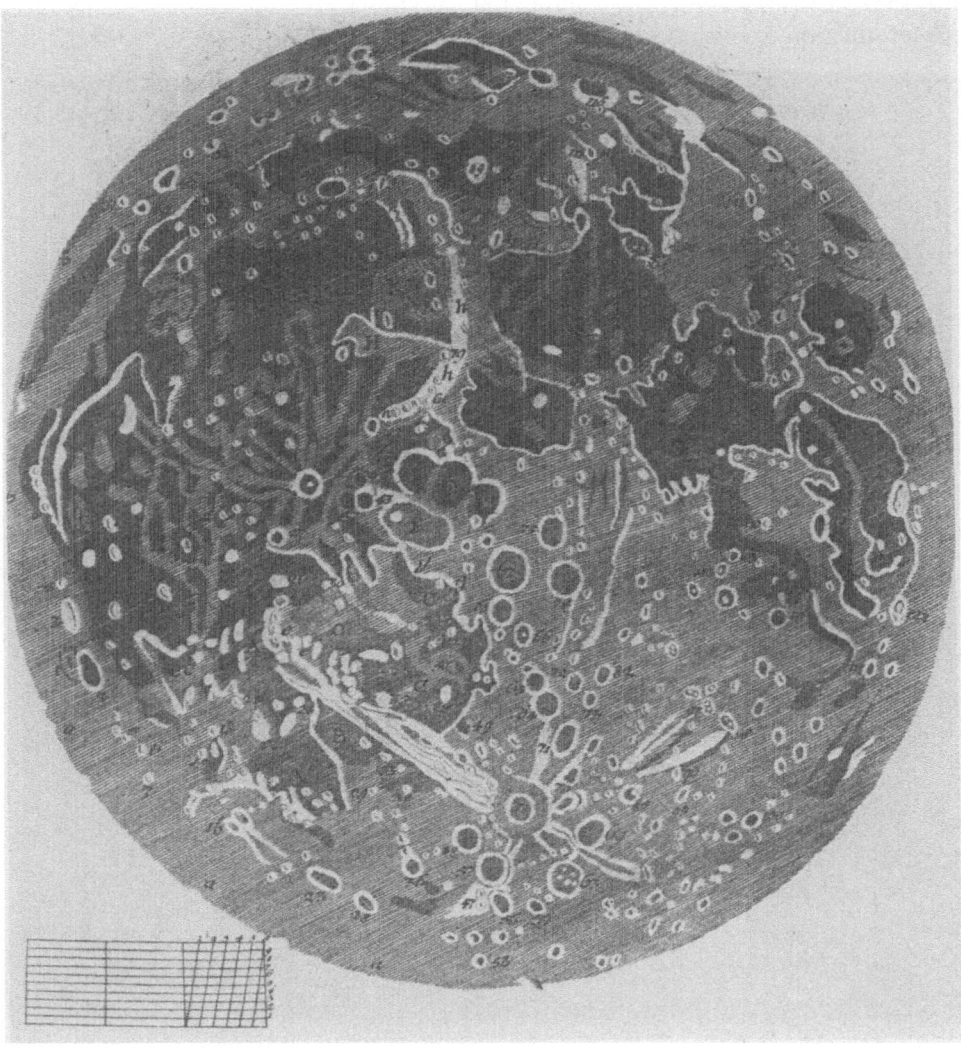

Fig. 1.21.   A map of the Moon by Maximilian Hell (1764).

the longitude $\lambda = 9°2'$ E and latitude $\beta = 14°34'$ N. With a glass micrometer Mayer measured the positions of 23 secondary control points, and estimated additional 65 points with reference to the former.

On this basis Mayer proceeded to construct two maps of the Moon in orthographic projection: one 19 cm, the other 45 cm in diam. A more detailed map of 25 sections was planned, but never completed. In 1751, Mayer became professor of mathematics at the University of Göttingen, where in the remaining years of his short life he made important contributions to the studies of the Moon's motion, but none more to lunar cartography. In point of fact, neither of Mayer's maps of the Moon appeared during his lifetime. The 19-cm map was eventually published by Georg Christian Lichtenberg in his *Tobiae Mayeri Opera Inedita* (Göttingen, 1775); and the 45-cm map not until 1881 by Ernst Klinkerfuess – more than a century after the death of its author. Its reproduction is given on the accompanying Figure 1.22.

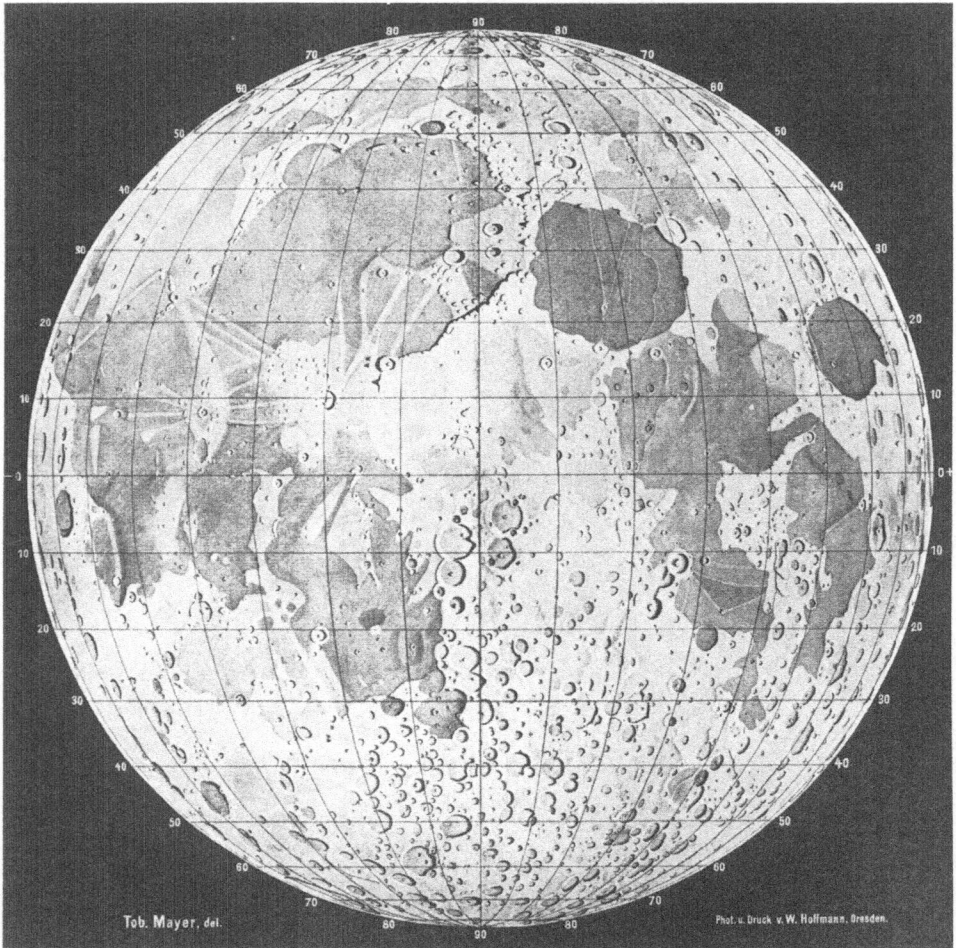

Fig. 1.22.   A map of the Moon based on the observations by Tobias Mayer around 1750, and published by Klinkerfuess in 1881 (North on top).

The principal innovation of this map is, of course, its equatorial net of coordinates in orthographic projection – representing the view of the near-side lunar hemisphere as seen from an infinite distance. Mayer's critical sense transpires also from a selection of the details recorded on his map: its comparison with modern work shows no important feature resolvable by the telescope at Mayer's disposal to be missing from the record; and the latter left no room for any poetical license on the part of the engraver. As such, Tobias Mayer became not only the first modern selenographer of the world, but also the founder of the German school of selenography which in the century to come 'took' the Moon away from the French and the Italians, and which included Schröter, Lohrmann, Mädler, Schmidt, and Fauth.

Tobias Mayer's work around 1750 – following by a century the work of Langren and Hevelius – represents a veritable landmark in the history of selenography which has since become a more exact branch of lunar studies. A year before Mayer's smaller map was published, Johann Heinrich Lambert (1728–1777) issued one of his own (Berlin, 1774) which represented, however, little or no advance over Mayer. A much more important contributor to selenography became Johann Hieronymus Schröter (1745–1816), since 1781 owner of a private observatory at Lilienthal in northern Germany equipped with $4\frac{3}{4}$ and 6-in. telescopes made by William Herschel, later to be replaced by instruments of diameters up to $18\frac{1}{2}$ in.

Between 1781–1797, Schröter carried out countless observations of the Moon, only a part of which was ever published. His original plans were to construct a map of the Moon 116 cm in diam – larger than Mayer's – and (an essential innovation) to determine the altitudes of lunar mountains from the length of the shadows which these mountains cast on the surrounding landscape in oblique solar illumination. H. W. M. Olbers (1758–1840), one of the leading theoretical astronomers in Germany at that time, lent a hand to Schröter with the geometry of the problem; thus a method was born, of the modern version of which an account will be given in Chapter 5 of this book.

The large map on the construction of which Schröter embarked in the 1780's was never completed; but the results of his observations eventually appeared in two parts – as *Selenotopographische Fragmente zur genaueren Kenntniss der Mondoberfläche* – in Lilienthal (1791) and Göttingen (1802). With the publication of these volumes Schröter truly opened a new era of lunar studies – involving detailed scrutiny of individual surface features under varying angles of illumination. His draftsmanship (or, more probably, that of his engraver – Tischbein by name – whose services Schröter engaged between 1789–1791) left, perhaps, something to be desired (for a sample of his work, see Figure 1.23); but his aims had increasingly been shared by most selenographers who followed Schröter in the 19th and 20th centuries.

After the publication of the second volume of his *Fragmente*, Schröter's attention gradually turned away from the Moon to the comets; but by that time Schröter's years on the Earth were almost numbered. In 1813, his observatory at Lilienthal, with most of his books and instruments, were largely destroyed by the vandalism of Napoleonic armies; and the ageing Schröter did not survive its end by more than

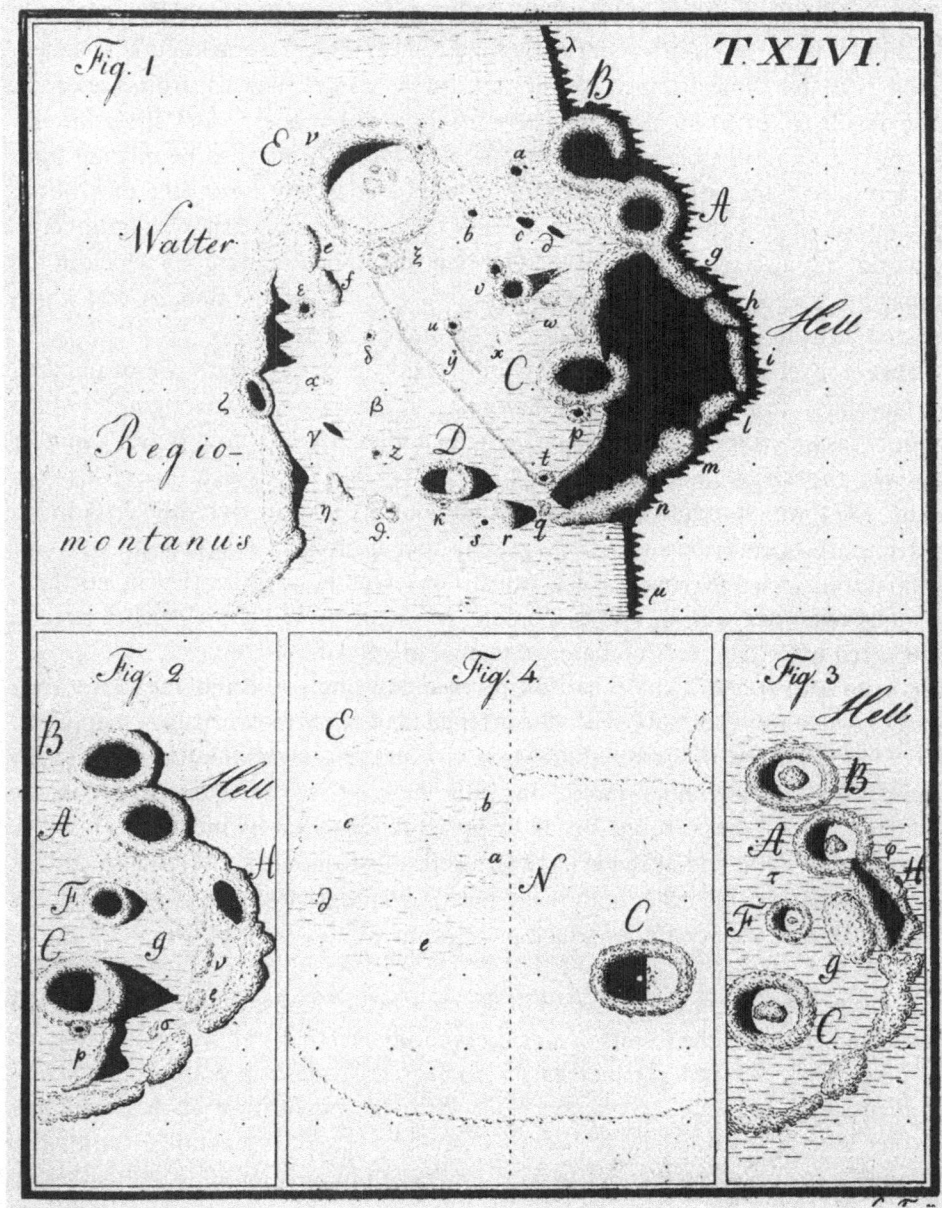

Fig. 1.23.   One of the drawings of the Moon by Schröter (1791), depicting the proximity of the craters Walter and Regiomontanus. For a more modern version of the same region, cf. the drawing by Philipp Fauth, reproduced on Figure 1.28.

three years – a fate to be re-enacted for another German selenographer of the same lineage, Philipp Fauth, in 1941.

As an original map-maker, Tobias Mayer did not find a worthy successor till in Wilhelm Gotthelf Lohrmann (1796–1840), a professional cartographer and surveyor of the Kingdom of Saxony. Armed with a 4.8-in. refractor of 1.8 m focal length, between 1822–1826 Lohrmann established the selenographic positions of 79 control points on the Moon by telescopic observations made with the aid of a filar micrometer, and (after a more ambitious start) divided his map of 97.5 cm in diam into 25 sections, the first form of which appeared under the title *Topographie der sichtbaren Mondoberfläche* (Dresden and Leipzig, 1824). This book, containing as it does an accurate description of the methods employed as well as of the results obtained, can be regarded as the first modern treatise on selenography.

Poor health delayed the completion of Lohrmann's map until 1836; and although a smaller general map (38.5 cm in diam) was issued by him in 1838, Lohrmann's premature passing two years later (like Mayer, Lohrmann too did not survive by much the age of 40) delayed the publication of the main map for many years. It was eventually edited by J. F. J. Schmidt, but did not actually see the light of the day until 38 years after Lohrmann's death (*Mondcharte von Wilhelm Gotthelf Lohrmann*, Leipzig 1878) – long after the publication of the famous map of the Moon by Beer and Mädler in 1834, and not before Schmidt's own great map of 1878. To give an example of Lohrmann's work, we reproduce on Figure 1.24 a part of his map of the region of the lunar Apennines from the 1878 edition of his work.

While Lohrmann was still occupied with his survey of the Kingdom of Saxony and of the lunar surface, a similar project even more extensive was initiated in Berlin, where Wilhelm Beer (1797–1850), a banker and amateur astronomer (brother of the poet Michael Beer, and half-brother of the composer Meyerbeer) joined forces with Johann Heinrich Mädler (1794–1874) to study the lunar surface at Beer's private observatory on the roof of his villa in Berlin. Around 1830 – when it became apparent that Lohrmann would not continue publication of his map, Beer and Mädler commenced with a construction of a new map of their own – a task which has occupied them for about five years.

In the course of this work, Mädler measured micrometrically, with Beer's Fraunhofer refractor of 9.4 cm free aperture, the positions of 105 control points on the lunar surface; but in contrast with Lohrmann, Mädler shifted the position of the fundamental point to the crater Mösting A which has continued to serve for this purpose on the Moon ever since. The resulting map of the Moon, 97.5 cm in diam (i.e. of the same size as Lohrmann's) was published in four sections under the title of *Mappa Selenographica totam Lunae hemisphaeram visibilem complectens* (Berlin, 1834–1836); and a sample of it is reproduced on the accompanying Figure 1.25.

The appearance of this Mappa Selenographica, accompanied by its companion volume *Der Mond, oder allgemeine vergleichende Selenographie* (Berlin, 1837) constituted another important milestone in the developement of selenographical literature. The accompanying volume (*Der Mond*) of over 400 pages contained, in addition

Fig. 1.24.   Region of the lunar Apennines as recorded on one section of a map by W. G. Lohrmann (Leipzig, 1878).

to a complete account of the methods of observation and reduction also a section on the history of lunar studies as well as a detailed description of the entire Moon's surface visible from the Earth. Moreover, the book summarized also the results of the determinations (by the shadow method) of the relative altitudes of 830 individual mountains, and of the diameters of 148 craters.

A second edition of the map by Beer and Mädler was published in 1877, and its reduced version (scaled down to 32 cm in size) in 1837. Although Beer's name is listed on both as that of co-author, it should be understood that his role in the joint under-taking has very largely been that of a benevolent patron whose support made the work possible; but the scientific results are actually due to Mädler.

The maps by Beer and Mädler remained unsurpassed in wealth of information for several decades. Their supremacy ended only in 1878, with the appearance of a much larger map by Johann Friedrich Julius Schmidt (1825–1884), perhaps the most prolific selenographer of the 19th century. In the course of work extending over 32 years (1842–1874) and carried out as several localities (mainly at Olomouc, Czechoslovakia and Athens, Greece), Schmidt accumulated the material for his monumental *Charte der Gebirge des Mondes* (Berlin, 1878), prepared on the scale 1:1 780 000 (correspond-

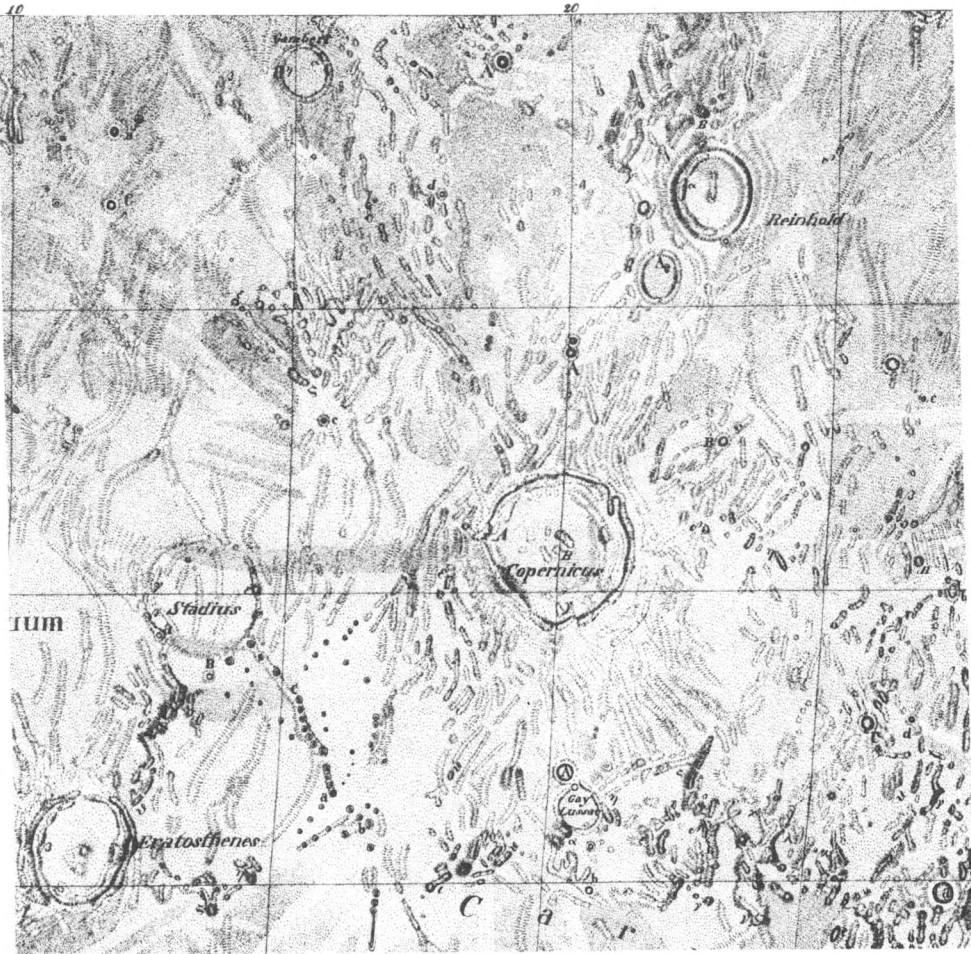

Fig. 1.25.   A section of the lunar map by Beer and Mädler (1837) of the region surrounding the crater Copernicus.

ing to a lunar diameter of 194.9 cm), more than twice as large as that adopted by Lohrmann or Beer and Mädler. The map itself was divided in 25 sections, each 40 by 40 cm in size; and a reproduction of one of these (depicting again the region of lunar Apennines) is shown on Figure 1.26.

    In constructing his map, Schmidt relied on the selenographic control points es-tablished by Lohrmann and Mädler. He measured, however, the relative altitudes of more than 3000 lunar mountains and eminences – all given in the *Ergänzungsband* (Berlin, 1878) to his map. The latter records almost 33 000 craters and other features on the Moon – in contrast with a little over 7000 craters recorded by Lohrmann, and some 7700 recorded by Beer and Mädler. As, however, Schmidt's work extended over more than 30 years (and was carried out with no less than nine different telescopes),

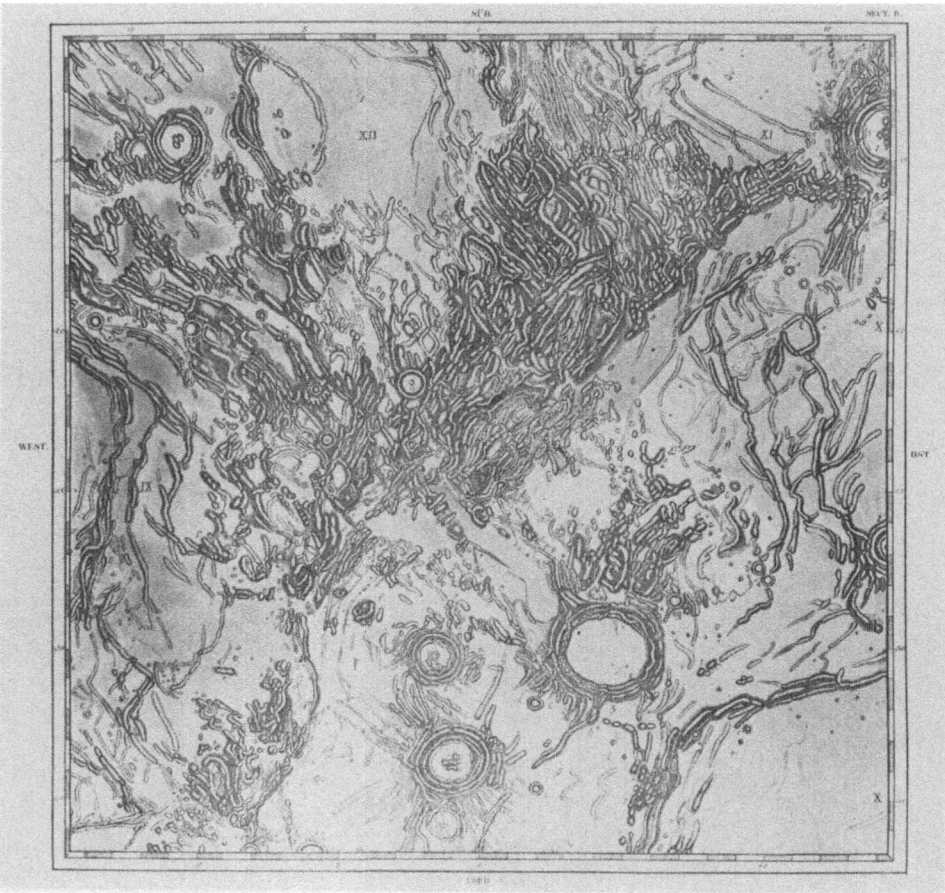

Fig. 1.26.   One of the 25 sections of the lunar map by J. F. J. Schmidt (Berlin, 1878) representing the lunar Apennines and adjacent parts of Mare Imbrium.

his data lack to some extent the homogeneity attained by some of his predecessors. Thus, Miss Blagg (1929) noted that ... 'Schmidt made a great many height measures on the Moon, and in the book accompanying his lunar maps he gives them combined with those of Beer and Mädler when possible. But, unfortunately, he made some mistakes of identification, and combined their measures of one height with his own of another. He also occasionally mixed his own measures ....' Therefore, Schmidt's map – in spite of its size and scale – does not represent as reliable a source of seleno-graphic data as was true of the work of some of his predecessors.

Attempts to surpass Schmidt's map in size were made (under the auspices of the British Association for the Advancement of Science) by William Radcliff Birt (1804–1881) who undertook between 1864 and 1870 to produce a large map of the Moon 5 m in diam; but only four sheets of it were completed before Birt's death, and in 1882 the whole project was abandoned. Of subsequent contributions to seleno-graphic literature forthcoming from the British Isles during the second part of the 19th century, James Nasmyth (1808–1890) and James Carpenter (1840–1899) com-bined efforts to publish, in 1874, a book entitled, *The Moon, considered as a Planet, a World, and a Satellite*, containing (apart from a small map) a remarkable series of photographs of plaster models of the lunar surface based on telescopic drawings – one of which, depicting the twin craters Aristarchus and Herodotus (with Schröter's canyon) is reproduced on the accompanying Figure 1.27. Edmund Neison (1851–1940) published in 1876 another book with a description of the lunar surface under the title, *The Moon and the Condition and Configurations of its Surface*, accompanied by a map 60 cm in size. A somewhat smaller map (46 cm in diam) was published in 1895 by Thomas Gwyn Elger (1838–1897). On the other side of the Channel, Casimir Marie Gaudibert (1823–1901) published in 1887 a map of the Moon 63.5 cm in diam, based on telescopic drawings noted for their accuracy.

In the first half of the present century, efforts to construct lunar maps comparable with, or larger than, Schmidt's on the basis of visual telescopic observations (aided, in some cases, by visual transcripts of photographs) were made only three times: namely, by Goodacre, Wilkins and Fauth.

Walter Goodacre (1856–1938), for many years Director of the Lunar Section of the British Astronomical Association, published in 1910 in London a map of the Moon, originally drawn on a single sheet with a diam of 192.5 cm. and afterwards subdivided into 25 sections on a scale of 60 in. to the Moon's diam. Its control points (1433 in number) were taken from the contemporary work of Saunder (1907); and the ensemble published in a separate book *The Moon*, which appeared in London in 1931.

More recently, two other large maps of the Moon have been published by Hugh Percival Wilkins (1896–1960) and Philipp Johann Heinrich Fauth (1867–1941). In 1924, Wilkins published his first lunar map, 152.4 cm in diam, followed in 1930 by one 508 cm in size. This was to be followed, in post-war years by one 762 cm in diam; but the latter was never to be published on that scale, but only on a scale reduced to one-third in 1946; and, in book form, in 1955 (Faber and Faber, London). The selenographic data plotted on Wilkins's maps are, however, rather heterogeneous, and the accuracy of its coordinate systems as well as workmanship in detail left much to be desired.

While the relatively large scales of Wilkins's maps have scarcely proved to be an unmixed blessing, the same cannot be said of the maps of the Moon which we owe to Philipp Johann Heinrich Fauth (1867–1941), the last of the great visual observers of the Moon, whose maps proved to be veritable works of art; an example of his execution being shown on the accompanying Figure 1.28. The first one, published in

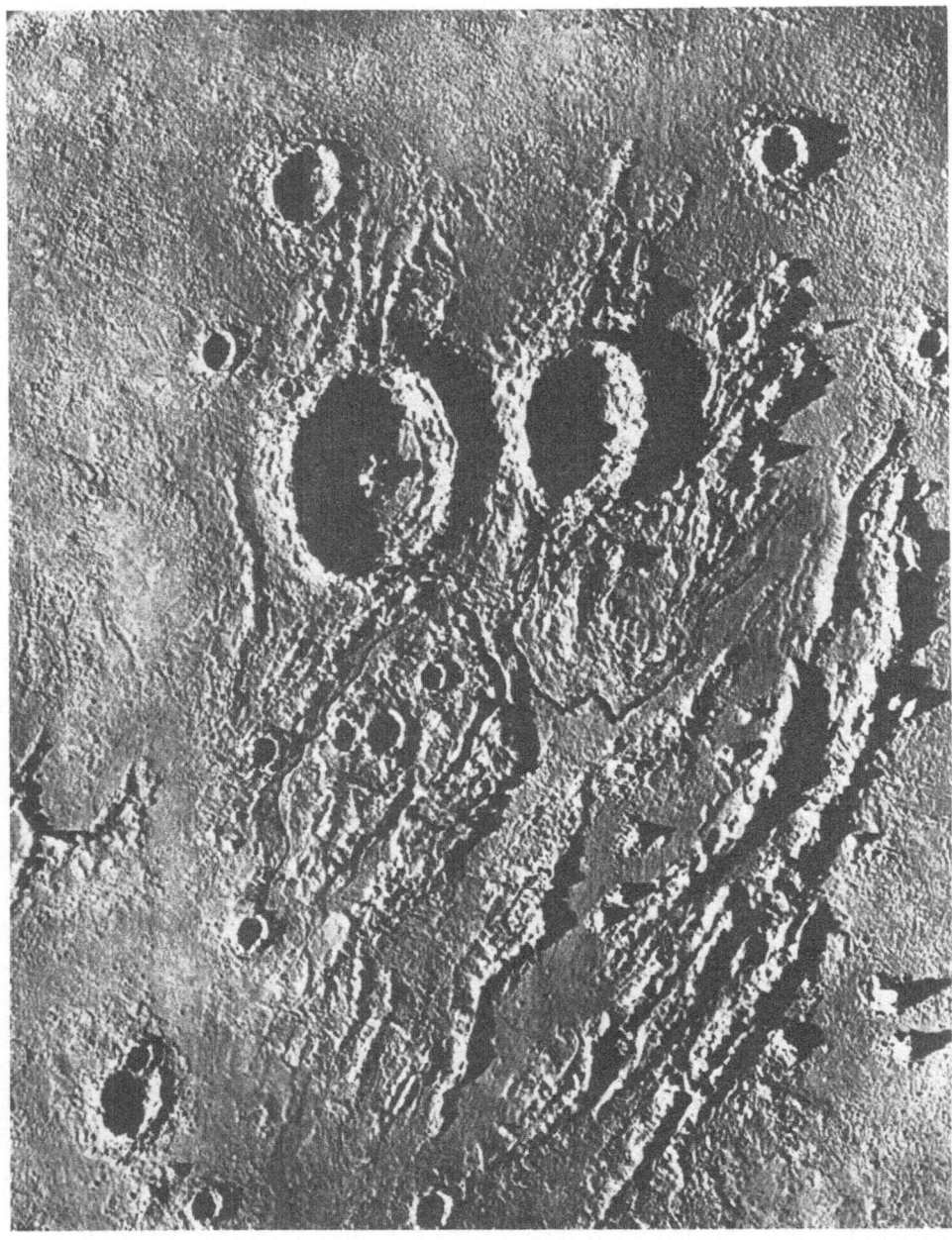

Fig. 1.27.   Photograph of a plaster model of the region of the lunar craters Aristarchus and Herodotus, by Nasmyth and Carpenter (1874).

1894, covered only 25 selected lunar regions. The second (in 1932) contained likewise only 16 large-scale maps. Fauth's great map of the Moon – 342 cm in diam – commenced to appear in 1932, but only five sections were completed by the time of its author's death. His son, Hermann Fauth, completed, however, the remaining 17 sheets from pencil drawings left by his father, and published the complete map in 1964. In 1934, Philipp Fauth published also a smaller map of the Moon, 87 cm in diam,

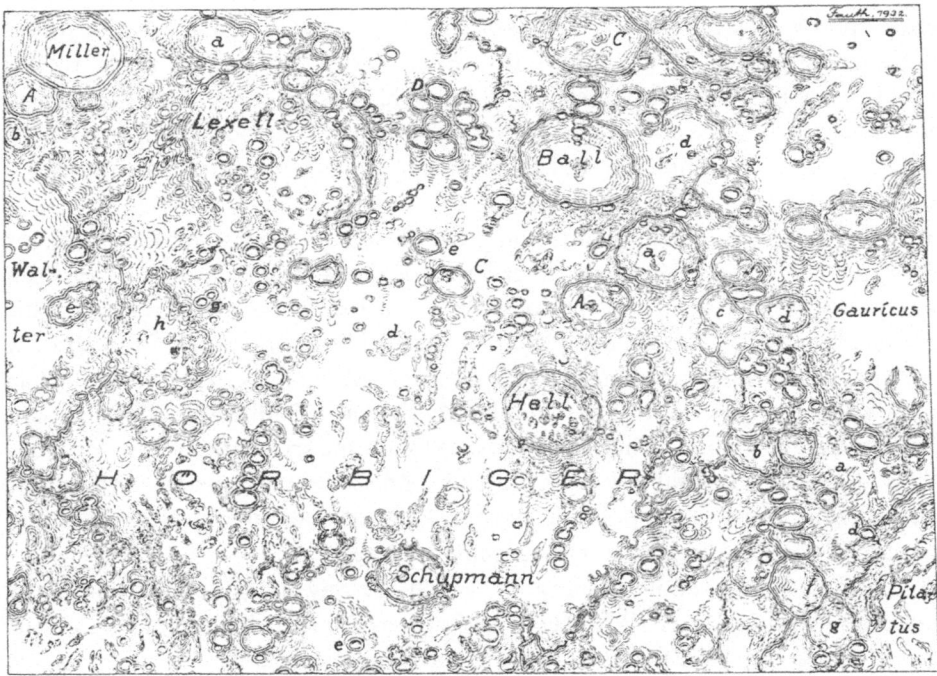

Fig. 1.28.   A map of the south-central portion of the lunar disc, prepared by Ph. Fauth in 1932, and published (posthumously) in 1964. The Hell plain designated (privately) by Fauth on his chart as 'Hörbiger' has since been renamed Deslandres; and the low-walled crater labelled 'Schupmann' is more generally known as Hell B.

drawn in 6 sheets to serve as a guide to his personal nomenclature of lunar formations in his book *Unser Mond*.

Of smaller maps of the Moon which appeared in the intervening years, mention should be made of two maps prepared by Karel Anděl (1884–1947), and published in Prague in 1926. These maps, 61 cm in diam, represent visual transcripts of lunar negatives taken with the 36-in. Lick refractor which remained in Prague after Weinek (see later in this chapter): one map represents the artist's view of a plastic relief of the lunar surface with the Sun about 30° above the horizon; while its companion (without relief) served as a carrier of nomenclature. Somewhat later, Felix Chemla Lamèch (1894–1962), published two somewhat less successful maps of the Moon: one in 1927 (45.7 cm in diam), and a somewhat larger one (61.5 cm) in 1934.

Lastly, mention should be made of a small *Atlas of the Moon*, published in London in 1935 by Mary Adela Blagg (1858–1944) and William Henry Wesley (1841–1922) under the auspices of the International Astronomical Union. The maps contained in this Atlas cover the whole face of the Moon divided into 14 sections. The four inner sections were drawn by Wesley between 1911–1912; while the ten outer sections, covering the limb regions on a different scale, were completed by Miss Blagg in 1922. The main purpose of these maps was to provide a basis to which Miss Blagg, in collaboration with Karl Müller (1866–1942) could refer a catalogue of the lunar nomenclature approved by the IAU in 1932: Vol. I of the 'Named Lunar Formations' constitutes a catalogue by Blagg and Müller; while Vol. II contains the Blagg-Wesley maps. In assessing the value of these maps, it should be noted that neither Miss Blagg, nor Wesley made any original observations themselves; therefore, their maps constitute second-hand compilations of topographic data taken from other sources.

The main reason why Philipp Fauth will probably remain the last author of a major map of the Moon based on visual observations is the fact that – as it has happened already in so many other branches of astronomical science – visual selenographic work has by now been almost completely superseded by photography. In 1839, a few years after the world was astounded by the notorious Moon hoax in the *New York Sun* on the alleged discoveries of Sir John Herschel (then in South Africa) of lunar inhabitants and their works; while some chose to follow the speculations by F. P. Gruithuisen about the inhabitability of our satellite, a discovery was made which, in due course, became primarily responsible for the realization that lunar surface is dead and immutable: namely, that of the photographic process.

The history of lunar photography goes back all the way almost to the cradle of the photographic process itself. When L. J. M. Daguerre, one or the originators of this process, devised in 1839 the way of copying photographic negatives on paper, he was encouraged by François Arago, then director of the Paris Observatory, to attempt a photograph* of the Moon, in order to discover whether or not the light of the Moon was chemically actinic. This proved indeed to be the case; though otherwise the experiment was a failure, for Daguerre's plate of the Moon showed no distinguishable detail on its face.

The failure did not, however, deter J. W. Draper (1800–1882) from repeating promptly Daguerre's experiment with improved means. He realised that the main cause of it was the fact that Daguerre underexposed his plates; and in order to avoid this, Draper exposed, in 1840, a 25-mm image of the Moon formed in the focus of his 12-in mirror of f/10 focal ratio for 20 min. This time the outcome proved to be a success, which was further improved by W. C. Bond (1789–1859), the first director of the Harvard College Observatory, who (working together with J. A. Whipple) used for this purpose the 15-in. f/20 Merz refractor then newly installed at Harvard. By 1850, these investigators were able to obtain lunar photographs exposed in less than

---

* It is of interest to note that the term 'photography' seems to have been first applied to the contemporary daguerrotypes by the selenographer J. H. Mädler.

1 min, which were capable of enlargement and showed details of all principal features of the surface of our satellite.

The next forward step in lunar photography was taken by Warren de la Rue (1815–1889), who was the first to use the collodium plates exposed in the focus of a 12-in. reflector of 305 cm focal length. His negatives, taken in the years 1852–1857, were sufficiently sharp to stand considerable enlargement, and can be regarded as true forerunners of the splendid series of photographs secured later by various large telescopes of the world. In 1854, improved photographs of the Moon were submitted at the meetings of the British Association for the Advancement of Science by Sir William Crookes and J. B. Reade. Crookes used a refractor of 203 mm in aperture and 390 cm focal length – while Reade took his photographs with the aid of a reflector of 60 cm diam and $23\frac{1}{2}$ m focal length; the telescope itself was stationary during exposures that lasted several seconds, and the guiding was done on the plateholder. In 1857, Thomas Grubb published many photographs of the Moon, taken with a refractor of 32 cm aperture and 610 cm focal length (guided likewise on the plateholder); and in 1858, L. M. Rutherfurd (1816–1892) secured the first stereoscopic pair of lunar photographs in New York. Shortly thereafter, further contributions to lunar photography were made in the U.S.A. by such well-known astronomers as Father Angelo Secchi (1818–1878) then in Georgetown, and Benjamin Apthorp Gould (1824–1896). Some of the best lunar photographs taken by Draper (1863) or Rutherfurd (1873) can be seen on the accompanying Figures 1.29 and 1.30.

All these investigators can be regarded as fathers of lunar photography near the middle of the 19th century. With the exception of Reade, they all worked with telescopes 8–15 in. of aperture; and the entire subsequent development of the subject was inseparably connected with the construction and use of the telescopes of larger resolving power, together with a gradual improvement of the photographic material and replacement of the collodium plates by dry bromosilver emulsions.

As a culmination of this effort in the 19th century, its last decade witnessed the publication of two great photographic atlases of the Moon, based on photographs secured at the Lick and Paris Observatories. Following the erection of the 36-in. refractor of 17.34 m focal distance at Lick Observatory in 1888, Edward S. Holden (1846–1914), the observatory's first director, used this excellent instrument for an extensive program of lunar photography; and a part of the results appeared under the title of the *Lick Observatory Atlas of the Moon* in 1896–1897. The enlargements reproduced in this Atlas correspond to a diameter of the Moon of 97.45 cm (i.e. to a scale of 1:3 547 000). The actual size of the illustration is 23 × 32 cm; and 1 mm on the prints corresponds to 3567 m on the lunar surface.

In subsequent years, Holden lent a considerable number of the Lick negatives to Ladislav Weinek (1848–1913), director of the Prague Observatory, who embarked on their systematic enlargement and printing. On the original Lick negatives, the diameter of the Moon's image oscillated between 12.4 cm at apogee to 13.9 cm at perigee. Weinek enlarged these 23.77 times, to make the diameter of the Moon correspond to 296 cm at apogee and 330 at perigee – so that 1 mm on the enlarge-

Fig. 1.29.   A photograph of the Moon taken by Henry Draper on September 3, 1863 with his $15\frac{1}{2}$-in. telescope. Note in the left-hand corner an inscription by the author to Sir John Herschel (original in possession of the Royal Astronomical Society in London).

Fig. 1.30.   A photograph of the Moon by L. M. Rutherfurd of Columbia University, New York, taken
on January 4, 1873 (reproduced from the original).

ments corresponds to 1115 m on the lunar surface (i.e. to a scale of 1:1115000).
Apart from a total of 114 Lick negatives received from Holden, Weinek also received
from Paris negatives from M. Loewy, which he enlarged 23–26 times to correspond
to a lunar diameter of 396 cm (i.e., to a scale of 1:877500).

The results of this work were published by Weinek under the title *Photographischer Mond-Atlas, wahrnehmlich auf Grund von focalen Negativen der Lick-Sternwarte in Maasstabe eines Monddurchmessers von 10 Fuss* (Prague, 1899). Originally Weinek intended to publish two volumes of his atlas, each consisting of 200 plates with the image size of 24.5 × 29.5 cm. As the *Atlas* is so arranged that each object is reproduced as it appears at the time of the lunar sunrise and sunset, its first 10 fascicles of 20 leaves each, which appeared between November 1897 and November 1900, recorded only 100 different sections of the lunar surface. The position and orientation of each plate was specified by the selenographic coordinates of its center, as well as by the selenographic longitude of the terminator.

That the second half of Weinek's *Atlas* as originally planned never appeared was probably due to its high price: for although its publication was subsidized by the Vienna Academy of Sciences as well as by Baron Rothschild and Miss Catherine Wolfe Bruce, it was offered commercially for sale at 100 Gulden – a sum equivalent to at least a monthly income of the highest-paid astronomers of that day. This effectively prevented adequate marketing and rendered Weinek's *Atlas* a rarity which only well-endowed libraries could afford to possess.

Moreover, the almost 24-fold enlargement adopted by Weinek for his work did not increase proportionally the amount of information discernible on the individual prints. In order to demonstrate this, we reproduce on Figure 1.31 a comparison of two

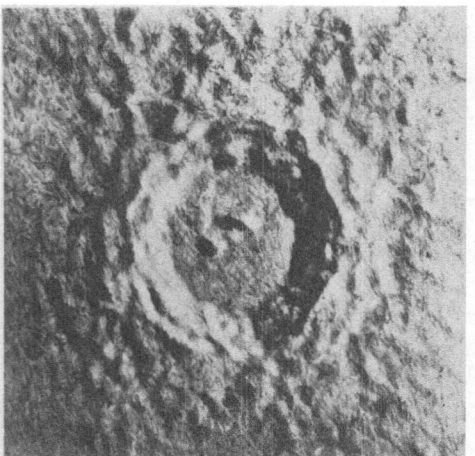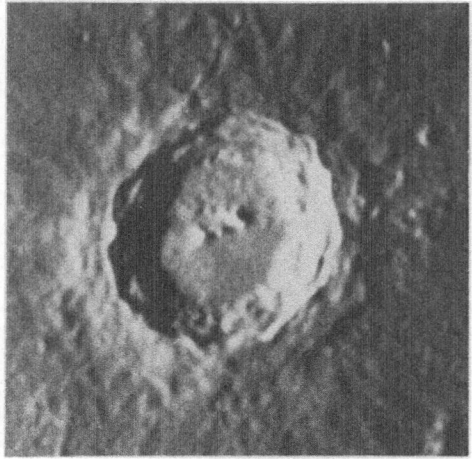

Fig. 1.31. A comparison of a visual transcription of the Lick photographs of the crater Copernicus by L. Weinek (1899), left, with a recent Pic-du-Midi photograph of the same formation on the right. The comparison reflects a vast improvement in photographic materials since Weinek's time.

photographs of the same region of the Moon (i.e., the crater Copernicus): Weinek's enlargement of a Lick negative (left) with a recent Pic du Midi photograph of the same formation and on the same scale (right). Both photographs were obtained on sites renowned for first-class seeing; but in spite of the fact that the photograph on the left was taken with a 36-in. objective, while only a 24-in. was used to secure the one on the right, the greatly superior quality of the latter reflects a vast improvement in photographic material in use since the days of Holden and Weinek. Nevertheless, such inordinate enlargements of these and other early photographs were used by other able selenographers – such as J. N. Krieger (1865–1902) – as a general background for detailed visual topography of individual lunar regions. His work, *J. N. Krieger's Mond-Atlas* (Wien, 1912), of which we reproduce Figure 1.32 as an illustration, was published subsequently by R. König (1865–1927).

Apart from the Holden and Weinek Atlases based on the Lick photographs of the Moon, the last years of the 19th century witnessed the commencement of another great contribution to selenographic literature: namely, the *Atlas Photographique de la Lune*, by M. Loewy and P. Puiseux, Observatoire de Paris, part 1–10, plates 1–80, 1896–1909. The negatives at the basis of this Atlas were secured by its authors and their collaborators at the Observatoire National in Paris with the aid of the 'grand equatoreal coudé' of 60 cm free aperture and 18 m focal length. The size of the lunar image in the focus of this refractor oscillated between 17.3 cm to 15.5 cm from perigee to apogee; and 80 enlargements of such negatives (unfortunately, not all on the same scale) constitute an atlas which remained the standard work on its subject for several decades.

As was the case with Weinek's *Atlas*, the physical size of the Paris *Atlas* make it likewise not easily accessible to every interested user; and, for this reason, several reduced editions of this standard work appeared in many countries. Thus M.C. Le Morvan published such an abridged edition in his *Carte photographique et systematique de la Lune* (Paris, 1914–1921) in 2 volumes (each in 4 parts), containing enlargements which correspond to a size of the lunar disc between 90 and 120 cm; the size of the individual prints being 25.5 × 32 cm. A Belgian edition of the Paris Atlas appeared in Brussels between 1899 and 1912 under the title of *Atlas lunaire, publié par la Société Belge d'Astronomie, reproduisant à une echelle reduite 2/5 les agrandissements photographiques de M. M. Loewy et P. Puiseux*. A Spanish edition by E. J. Thost, entitled *Resume del Atlas fotografico de la Luna del Observatorio Nacional de Paris*, appeared in Tarragona in 1922 (60 plates 15 × 21 cm in size); and was reproduced also in German in Stuttgart.

Apart from the Paris Atlas and its various editions which dominated the field in the first half of the present century, we may mention W. H. Pickering's *Photographic Atlas of the Moon*, which appeared as Vol. 51 of the *Annals of the Harvard College Observatory* in Cambridge, Mass., 1903, and contained on 88 plates reproductions (unenlarged) of photographs taken by Pickering with a 12-in. telescope of 41.5 m focal length at Mandeville, Jamaica. Because of relatively long exposure required by so large a focal ratio, the majority of Pickering's photofraphs lack the contrast and definition of the Paris plates; the merit of his work resting on the fact that each region of the Moon was photographed at five different elevations of the Sun.

Much more recently, a photographic atlas of the Moon on a similar scale was published by Sh. Miyamoto and M. Matsui as No. 95 of the *Contributions from Kwasan Observatory*, Kyoto, in 1960. The photographs reproduced on 85 plates were

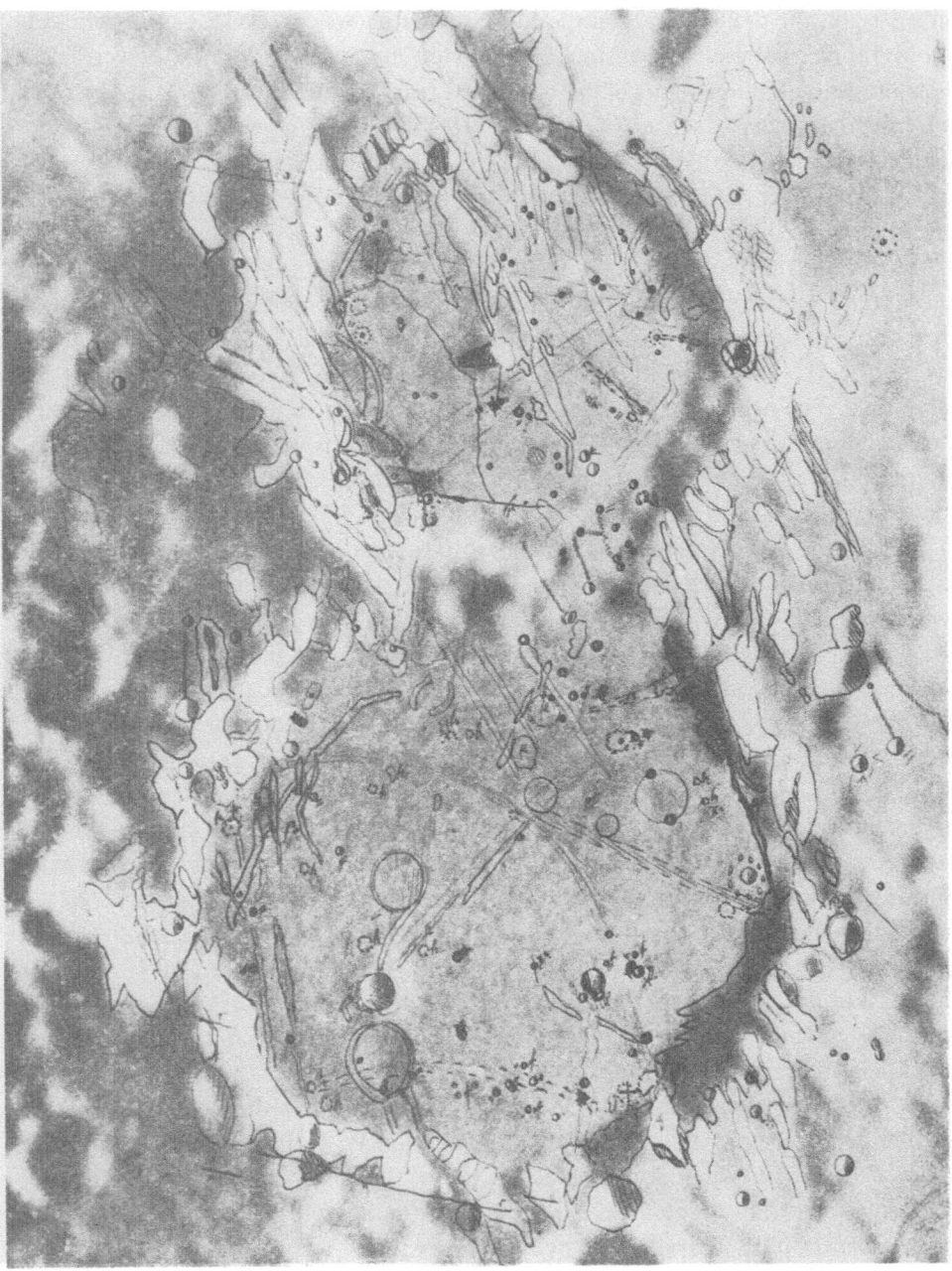

Fig. 1.32.   An example of Krieger's use of photographic enlargements as a basis for subsequent visual
work in the region of the craters Ptolemaeus and Alphonsus.

secured with the aid of a 12-in. refractor, operating at an effective focal length of 20.70 m.

Until 1960, the Paris photographic Atlas of the Moon by Loewy and Puiseux remained the most extensive work of its kind in existence. That year it was, however, superseded by a new *Photographic Lunar Atlas*, edited by G. P. Kuiper in collaboration with D. W. G. Arthur, E. Moore, J. W. Tapscott and E. A. Whitaker (Chicago, 1960). This atlas contains 281 illustrations, of which 212 exhibit 44 lunar regions under 4–5 different conditions of illumination by the Sun, on the scale of 1:370000. All these illustrations are based on enlargements of the photographs secured by many different astronomers at the Lick, Mt. Wilson, McDonald, Pic du Midi, and Yerkes Observatories, and collected by the compilers. Moreover, a subsequent appendix to this atlas (*Orthographic Atlas of the Moon*, compiled by D. W. G. Arthur and E. A. Whitaker in 1962) contains overprints of lunar co-ordinates superimposed over photographs of a major part of the visible lunar hemisphere; while a second appendix (in the form of *Rectified Lunar Atlas*, by E. A. Whitaker, G. P. Kuiper, W. K. Hartmann and L. H. Spradley, 1964) contains reproductions of photographs of the entire visible lunar hemisphere rectified by projection on a sphere.

All major lunar mapping programmes of the last decade will be described more fully in the second part of this book. In conclusion of the present chapter we wish only to note the appearance of some additional atlases of the Moon since 1965, which still remain of general interest – even though they may be superseded in detail by concurrent larger efforts based increasingly on space-borne data from closer proximity of our satellite.

Thus the year of 1965 – when the hard-landing Rangers impacted on the Moon – witnessed the (somewhat belated) publication of a *Photographic Atlas of the Moon* by Z. Kopal, J. Klepešta and T. W. Rackham (Academic Press, New York and London), based wholly on photographs secured between 1961–1962 by its authors with the aid of the 24-in. refractor of the Observatoire du Pic du Midi* in France. This atlas differs from its predecessors in so far as only terminator photography (both sunrise and sunset) has been included in the main part of the Atlas, consisting of 197 separate plates; and an accompanying skeleton map of the Moon's front side by A. Rükl contains a complete key to lunar nomenclature accepted up to that time.

In 1967, G. P. Kuiper with his associates (E. A. Whitaker, R. G. Strom, J. W. Fountain and S. M. Larson) brought out a *Consolidated Lunar Atlas* (as Supplement Nos. 3 and 4 to the USAF *Photographic Lunar Atlas*), containing 192 plates of reproductions (11 × 14 in. in size) of lunar photographs taken with the 61-in. NASA telescope of Catalina Observatory (Part I), and 34 plates of photographs taken with the 61-in. astrometric reflector of the U.S. Naval Observatory at Flagstaff. This atlas will probably remain the last major undertaking of this kind based exclusively on

---

* Its objective being a twin of the lens of the grand equatoreal coudé at Paris Observatory, with which Loewy and Puiseux secured the photographs for their well-known Atlas of the Moon between 1896 and 1909. In 1942 it was re-mounted by Lyot at Pic du Midi, where it remained in operation, and in the service of lunar photography, until the summer of 1971 when its optics was recalled to Paris; and the Lyot dome where it was housed has become the place of purely solar research.

photographs taken from the Earth. The *New Photographic Atlas of the Moon* by Z. Kopal (Taplinger Publ. Co., New York, 1971) contains only about one-half of ground-based photographs of the Moon (taken very largely with the 43-in. reflector of the Observatoire du Pic du Midi); the balance of its 214 plates (9 × 12 in. in size) having been secured by different types of spacecraft.

Of these, the primacy before the summer of 1969 belonged to the lunar orbiters of the years 1966–1967, which first brought high-resolution optics to the proximity of our satellite. Following a preliminary NASA Special Publication No. 200, entitled *The Moon as Viewed by Lunar Orbiter* (Washington, D.C., 1970) by L. J. Kosofsky and Farouk El-Baz, an impressive *Lunar Orbiter Photographic Atlas of the Moon* (NASA SP-206, Washington, D.C., 1971), prepared by D. E. Bowker and J. K. Hughes of the NASA Langley Research Center, contains excellent reproductions of 675 plates taken by the U.S. orbiting satellites 1–5 in 1966–1967, $10 \times 13\frac{1}{2}$ in. in size; with full particulars about each plate. This, together with the accompanying *Gazeteer of the Near Side of the Moon* (NASA Sp-241; Washington, D.C., 1971) prepared by the Mapping Sciences Laboratory of NASA Manned Spacecraft Center in Houston, constitute important additions to recent selenographic literature, opening up the wealth of space-borne photography to a wider community of interested readers.

Last but not least, the year 1972 witnessed the appearance of another valuable contribution to selenographical literature in the form of Antonín Rükl's *Maps of Lunar Hemispheres*, giving – for the first time in the history of lunar mapping – the views of the Moon as seen from six cardinal directions in space. This publication, which appeared as Vol. 33 of the *Astrophysics and Space Science Library* (published by D. Reidel Publ. Co., Dordrecht, Holland) consists of six charts, $21 \times 24\frac{1}{2}$ in. in size, on which the Moon is recorded in Lambert's azimuthal (equi-areal) projection as it would appear to an external observer from six mutually perpendicular directions in space. Map No. 1 (the near side, visible from the Earth) previously accompanied the Kopal-Klepešta-Rackham *Atlas* of 1965, as well as Kopal's *Introduction to the Study of the Moon* (D. Reidel Publ. Co., Dordrecht, 1966); while Map No. 3 (the Moon's far side) appeared as a fold-out with Kopal's treatise on *The Moon* (D. Reidel Publ. Co., Dordrecht, 1969); but Rükl's Maps No. 2 and 4–6 are entirely new, and will no doubt be welcomed by all those who may want to follow in their imagination the lunar astronauts from their terrestrial armchairs.

On the more popular side – of lesser scientific significance, but of perhaps greater interest for the general public – we may note the appearance of Vincent de Callatay's *Atlas de la Lune* (De Visscher, Brussels, 1962; English edition by Macmillan and Co., London, 1964); as well as Dinsmore Alter's *Pictorial Guide to the Moon* (T. Y. Cowell Co., New York, 1963; second edition, 1967), followed by Alter's *Lunar Atlas* of 154 Plates, prepared by the Space Sciences Laboratory of the Space Division of North American Aviation, Inc. (Dover Publications, New York, 1964). The *Times Atlas of the Moon*, edited by H. A. G. Lewis (New York and London, 1969), consisting very largely of the reproductions of the LAC series of the ACIC maps of the Moon, belongs to the same category.

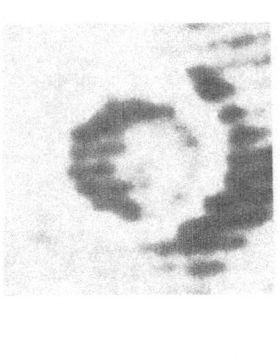

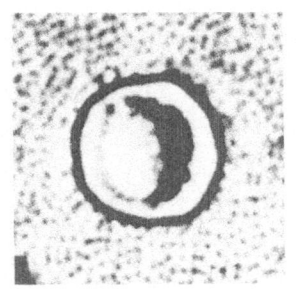

MELLAN (1637)   LANGREN (1645)   RHEITA (1645)

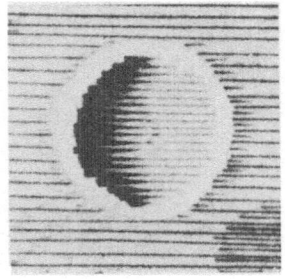

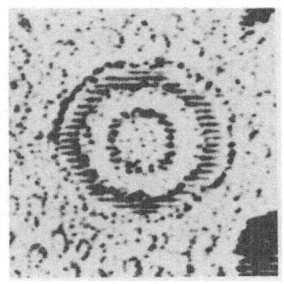

FONTANA (1646)   HEVELIUS (1647)   DIVINI (1649)

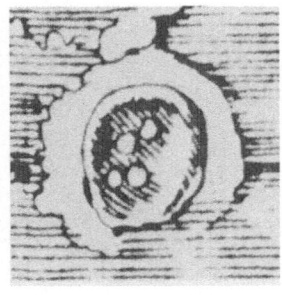

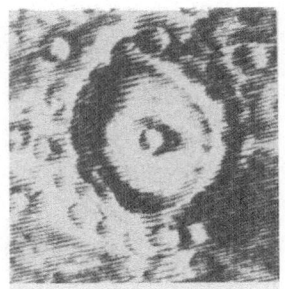

RICCIOLI (1651)   MONTANARI (1662)   CASSINI (1680)

Fig. 1.33.   Cartographic representations of the crater Copernicus on different lunar maps of the 17th
century referred to in the text (north on top; west to the left).

Fig. 1.34. Cartographic representations of the crater Copernicus on different lunar maps of the 18–19th centuries (north on top; west to the left).

GOODACRE (1910)   ANDĚL (1926)   FAUTH (1932)

BLAGG AND WESLEY (1935)   WILKINS (1951)   LAMĚCH (1957)

ACIC (1960)   ACIC (1964)   DIRECT PHOTOGRAPH

Fig. 1.35. Cartographic representations of the crater Copernicus on different lunar maps of the 20th century (north on top; west to the left).

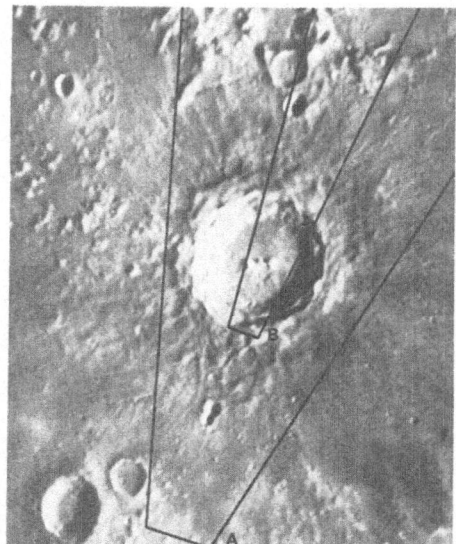

Fig. 1.36.   The crater Copernicus as photographed from the Earth (upper right), in comparison with views of the same formation as recorded by Orbiter 2 on 23 November, 1966, from closer proximity to its target. Photograph on the upper left shows an oblique view of this crater (with camera pointed 17° below the horizon) as seen through the Orbiter's wide-angle lens (from the position marked by A on the ground-based picture on the right): while the photograph below shows a high-resolution view of the interior of the crater from the vanrage point B. At the moment when this latter photograph was taken, the Orbiter was overflying the southern ramparts of Copernicus, at an altitude of 46 km. – The group of hills in the foreground of the photograph below constitutes the 'central mountain' of Copernicus.

The Russian contributions to the mapping of the Moon since 1959, based on photographic work of their spacecraft, will be dealt with in a separate chapter 10 in the latter part of this book.

The main aim of this introductory chapter for our monograph on lunar mapping has been to accompany the reader interested in the history of our subject from pretelescopic days to the advent of the space age; and thus set the stage on which the main developments have taken place in the past decade – developments which would have made selenographers of two or three generations before our time to gasp in awe. Nothing can, perhaps, telescope accelerating speed of these developments better than a glance at the accompanying Figures 1.33–1.35, at which we mounted (on the same scale, with north on top) graphical representations of the crater Copernicus as seen by the eyes of our predecessors – from Mellan in 1637 to the mid-20th century. The reader may, in this connection, compare also Figure 1.30 – before he turns to Figure 1.36 which will introduce to us this object of so many past studies as it was seen by the optical eyes of U.S. Lunar Orbiter 2 on 23 November 1966 in all its stark beauty. Can anyone contemplating this comparison doubt that, for the inquisitive human mind, the sky is no longer the limit?

References

Of the general literature of the past decade concerned with the subject of this chapter we may quote:

Kopal, Z.: 1962, 'Topography of the Moon', in *Physics and Astronomy of the Moon*, Academic Press, New York and London, Chapter 7.

Kopal, Z.: 1966, 'Mapping of the Moon', in *An Introduction to the Study of the Moon*, D. Reidel Publ. Co., Dordrecht, Holland, Chapter 15.

Kopal, Z.: 1969a, 'Mapping of the Moon', in *The Moon*, D. Reidel Publ. Co., Dordrecht, Holland, Chapter 15.

Kopal, Z.: 1969b, 'The Earliest Maps of the Moon', *Moon* **1**, 59–66.

Maffei, P.: 1963, 'Carte Lunari di ieri e di oggi', *L'Universo* **42**, No. 5–6, Istituto Geogr. Militare, Firenze, Italy.

Vyver, O. van de: 1971a, 'Original Sources of Some Early Lunar Maps', *J. Hist. Astron.* **2**, 86–97.

Vyver, O. van de: 1971b, 'Lunar Maps of the XVIIth Century', *Vatican Obs. Publ.* **1**, No. 2.

Of other references made in the text, which were not fully quoted, we should list:

Blagg, M. A.: 1929, *J. Brit. Astron. Assoc.* **39**, 328.

Houzeau, J. C.: 1882, in *Vade-Mecum de l'Astronome*, Bruxelles, pp. 550–561.

Humboldt, A. von: 1858, in *Kosmos*, Vol. 3, Stuttgart and Tübingen p. 544.

Mee, A.: 1908, *Knowledge* **5**, 280.

Rigaud, S. P.: 1833, Supplement to Bradley's *Miscellaneous Works*, Oxford, pp. 19–20.

Strout, E.: 1965, *J. Brit. Astron. Assoc.* **75**, 100.

Wolf, R.: 1890, *Handbuch der Astronomie, ihrer Geschichte und Literatur*, Vol. 1, Zürich, p. 100.

Zach, F. X. von: 1788, in *Berliner Astron. Jahrbuch* for that year.

CHAPTER 2

ROTATION AND LIBRATIONS OF THE MOON

A *conditio sine qua non* of the mapping of the Moon is the adoption of a definite system of *lunar coordinates*, to which the location of any point on the lunar surface could be uniquely referred. Before we can introduce these, however, it is necessary to lay down the ground on which such an adoption can be endowed with definite physical meaning: and the latter is inseparably connected with the problem of the generalized *rotation* of the lunar globe in space.

This rotation is also of paramount importance when we come to consider the inverse problem – that of determining the relative position of a point of given seleno-graphic coordinates on the apparent lunar disc as seen from a given vantage point – be it on the Earth's surface or aboard a lunar spacecraft – at any time. The apparent positions of such a point on the Moon are bound to *librate* with respect to an observer situated at a *finite* distance; and such librations must be taken into account in any determination of the lunar coordinates from an external vantage point. The aim of the present chapter will be to lay down the essential elements of the theory of lunar rotation and of the 'librations' of its apparent disc; and to postpone the definition of the lunar coordinates based upon it to the next chapter.

The revolution of the Moon around the Earth, and with the Earth around the Sun, are not the only motions performed by our satellite. As has been realized in the earliest days of lunar observations – from the fact that the Moon exhibits to us on Earth (almost) always the same face – it must also rotate with a uniform angular velocity about an axis fixed in space and inclined but little to its orbital plane.

Towards the end of the 17th century, the Italian-French astronomer Cassini, mentioned already on p. 21 in connection with his map, deduced from a long series of observations three empirical laws respecting lunar rotation, which bear his name, and can be expressed as follows:

(a) The Moon rotates eastward, about a fixed axis, with constant angular velocity and in a period to one sidereal month;

(b) The inclination I of the Moon's axis of rotation to the ecliptic remains constant; and

(c) The poles of the Moon's axis of rotation, of the ecliptic and of the lunar orbit lie in the same plane, and on one great circle, in that order (cf. Figure 2.1).

The angle i between the Moon's orbit and the ecliptic is known with great exactitude to be equal to $5°8'43''.4$ (Brown, 1919). The inclination I of the lunar equator to the ecliptic is known with somewhat lesser precision to amount to $I = 1°32'1'' \pm 7''$ (Koziel, 1967a, b). Hence, in accordance with Cassini's third law, the inclination of the Moon's equator to its orbital plane should be constant and equal to the sum of $I + i = 6°40'44''$. These facts entail, in turn, certain interesting consequences which manifest themselves to us as the *optical librations* of our satellite.

The oft-repeated statement that the Moon exhibits to the Earth always the same face on account of the synchronism prevailing between rotation and revolution is only approximately true, but cannot be exact for several reasons. The first is the fact

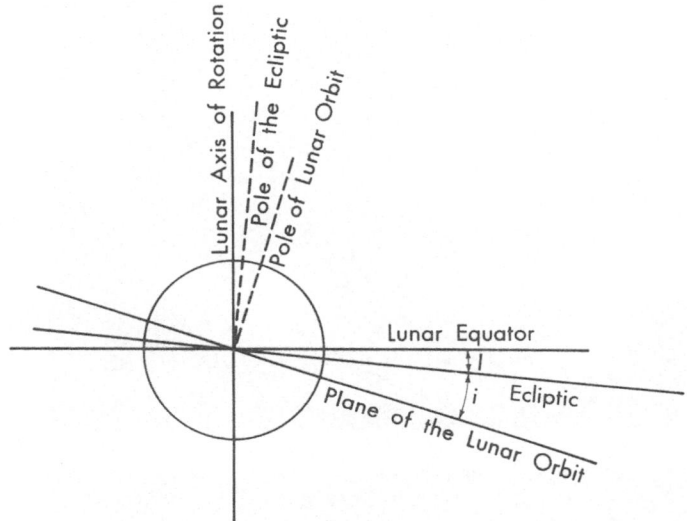

Fig. 2.1.   Relative orientation of Moon, Earth and of the Ecliptic.

that, if the axial rotation of the Moon is uniform (Cassini's first law), the angular velocity of its revolution in an elliptic orbit (varying as it does with inverse square of the radius vector, in accordance with Kepler's second law) is not; being sometimes ahead, and at other times behind, the rotational motion. The effect of this must be an alternating periodic angular displacement of lunar objects in longitude by as much as $7°57'$ about the centre of the Moon – a phenomenon known as the *optical libration in longitude* (cf. Figure 2.2).

Secondly, as the lunar axis of rotation is not perpendicular to the orbital plane, but its inclination deviates from 90° by $I + i$ (Cassini's second and third laws), we can see sometimes more of one polar region and, at other times, more of the other in the course of each month. This phenomenon gives rise to an *optical libration in latitude* by $6°51'$ (Figure 2.2). Again, when the Moon is rising for the observer on the Earth we look over its upper edge, seeing a little more of that part of the Moon than if we were observing it from the Earth's centre; and when the Moon is setting the converse is true. This *diurnal libration* (not of the Moon, to be sure, but of the observer) amounts to $\pi_\ell = 57'2''6$, and superposes upon all other librations to enable us to see considerably more than one-half of the lunar surface from the Earth. On the whole, not less than

Fig. 2.2.   A comparison of the two aspects of full-Moon face, as influenced by the optical librations of our satellite, and recorded on photographs taken with the 60-cm refractor of the Pic du Midi Observatory by M. Moutsoulas on August 30, 1966 at $20^h40^m15^s$ (libration constants, $l = +4.8$, $b = +6.2$) left and by G. L. Roberts on December 25, 1966 at $21^h46^m25^s$ (libration constants, $l = -5.4$, $b = -2.9$) right.

59% of the entire lunar globe can be seen at one time or another from the Earth; only 41% being permanently invisible, and an equal amount never disappearing; the remainder of 18% being alternately visible and invisible. In October 1959, the cameras aboard the Russian third space station (Luna 3) unveiled also the essential features of a major part (over 50%) of the far side of the Moon. The relative position of Luna 3 at the time of the photography was such that not all of the lunar far side was exposed to the camera, and approximately 13% of it remained completely uncharted till July 20, 1965, when another Russian space probe (Zond 3) succeeded in recording all but a small fraction of this remainder (Lipsky, 1965). The rest has since been virtually covered by the U.S. Lunar Orbiters of 1966–1967 and other spacecraft since that time; so that at present only a very small fraction of the lunar surface near the Moon's south pole (less than 1% of its entire surface) remains still unexplored.

The actual magnitude of lunar optical librations at any time is easy to ascertain from known elements of the Moon's motion. Let, in what follows, b and l denote the selenographic latitude and longitude of the Earth's centre (or, what is the same, the latitude and longitude of the apparent centre of the lunar disc as would be seen from the centre of the lunar disc as would be seen from the centre of the Earth). These (geocentric) *optical libration angles* can be evaluated from the equations

$$\cos b \cos(l + l_{\bullet} - \Omega + \Delta) = -\cos(\alpha_{\zeta} - \Omega') \cos \delta_{\zeta},$$
$$\cos b \sin(l + l_{\bullet} - \Omega + \Delta) = -\sin(\alpha_{\zeta} - \Omega') \cos \delta_{\zeta} \cos j - \sin \delta_{\zeta} \sin j, \qquad (2.1)$$
$$\sin b = \sin(\alpha_{\zeta} - \Omega') \cos \delta_{\zeta} \sin j - \sin \delta_{\zeta} \cos j,$$

in which α_{ζ}, δ_{ζ} denote the instantaneous right-ascension and declination of the Moon in the sky; and $l_{\bullet}$, the mean longitude of the Moon in its orbit. The angle Ω denotes the longitude of the ascending node of the lunar orbit in the ecliptic; Ω', the ascending node of the lunar equator in the plane of the terrestrial equator (both Ω and Ω' being measured from the vernal equinox). If, furthermore, ε denotes as usual the obliquity of the ecliptic, the angle j between the terrestrial and lunar equators is given by the equation

$$\cos j = \cos I \cos \varepsilon + \sin I \sin \varepsilon \sin \Omega, \qquad (2.2)$$

and in addition, the angles Ω and Ω' are related by

$$\sin \Omega' = -\sin I \csc j \sin \Omega; \qquad (2.3)$$

while Δ, the angular distance between the terrestrial equator and the ecliptic, measured along the lunar equator, is given by

$$\sin \Delta = -\sin \varepsilon \csc j \sin \Omega, \qquad (2.4)$$

Equations (2.1) define the angles b and l of lunar optical librations in terms of the Moon's equatorial coordinates $(\alpha_{\zeta}, \delta_{\zeta})$ and of the elements of lunar orbit. An alternative way to do so in terms of the Moon's ecliptical latitude β_{ζ} and longitude λ_{ζ} in the sky is offered by the equations

$$\cos b \cos(l + l_\zeta - \Omega) = \cos(\lambda_\zeta - \Omega) \cos \beta_\zeta,$$
$$\cos b \sin(l + l_\zeta - \Omega) = \sin..(\lambda_\zeta - \Omega) \cos \beta_\zeta \cos I - \sin \beta_\zeta \sin I, \qquad (2.5)$$
$$\sin b = -\sin(\lambda_\zeta - \Omega) \cos \beta_\zeta \sin I - \sin \beta_\zeta \cos I,$$

where λ_ζ and β_ζ related with α_ζ and δ_ζ by the well-known equations

$$\cos \beta_\zeta \cos \lambda_\zeta = \qquad\qquad + \cos \delta_\zeta \cos \alpha_\zeta,$$
$$\cos \beta_\zeta \sin \lambda_\zeta = \sin \delta_\zeta \sin \varepsilon + \cos \delta_\zeta \sin \alpha_\zeta \cos \varepsilon, \qquad (2.6)$$
$$\sin \beta_\zeta = \sin \delta_\zeta \cos \varepsilon - \cos \delta_\zeta \sin \alpha_\zeta \sin \varepsilon,$$

yielding

$$\tan \lambda_\zeta = \cos(N - \varepsilon) \sec N \tan \alpha_\zeta,$$
$$\tan \beta_\zeta = \tan(N - \varepsilon) \sin \lambda_\zeta, \qquad (2.7)$$

where we have abbreviated

$$\tan N = \tan \delta_\zeta \csc \alpha_\zeta. \qquad (2.8)$$

Before we proceed to solve Equations (2.5) for b and l, it is useful to introduce explicitly the angle i between the Moon's orbit and the ecliptic in our computations. From the spherical triangles outlined in Figure 2.3 it follows that

$$\sin \beta_\zeta = \sin \{ \tan^{-1} [\tan(\lambda_\zeta - \Omega) \sec i] \} \sin i \qquad (2.9)$$

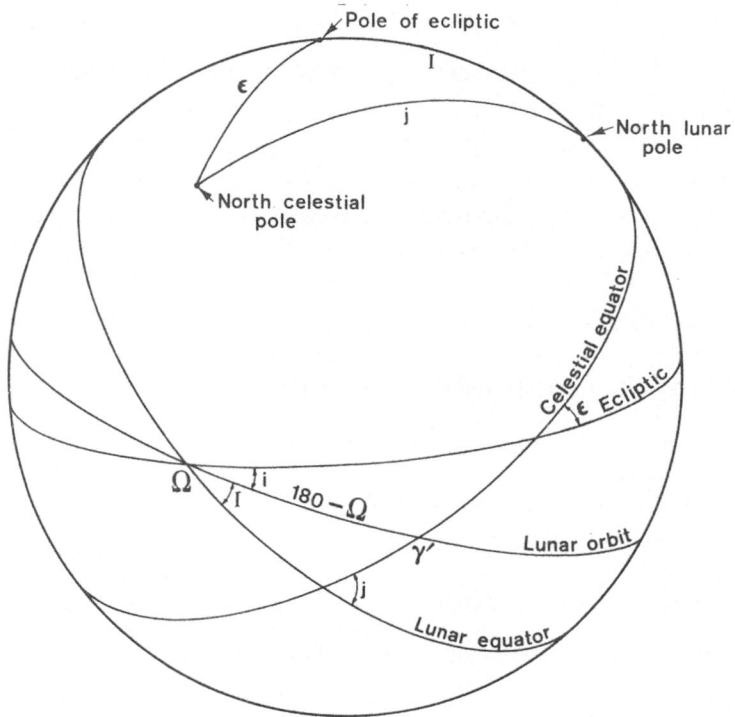

Fig. 2.3.   Selenocentric celestial sphere.

and

$$\cos\beta_{\langle}=\cos\left\{\tan^{-1}\left[\tan(\lambda_{\langle}-\Omega)\sec i\right]\right\}\sec(\lambda_{\langle}-\Omega),\tag{2.10}$$

which on insertion in (2.5) permit us to solve the latter equations for l and b in the form

$$l=\tan^{-1}\left\{\tan(\lambda_{\langle}-\Omega)\cos(I+i)\sec i\right\}-(l_{\langle}-\Omega)\tag{2.11}$$

and

$$b=\tan^{-1}\left\{\sin(l_{\langle}-\Omega+l)\tan(I+i)\right\},\tag{2.12}$$

respectively.

The values of b and l given by the foregoing equations represent the *geocentric* librations of our satellite. Their *topocentric* values b', l' referred to the actual position of the observer can, however, be obtained from the same systems of Equations (2.1) or (2.5), in which *the geocentric coordinates* $\alpha_{\langle}$, $\delta_{\langle}$ *or* $\lambda_{\langle}$, $\beta_{\langle}$ *of the Moon have been replaced by their topocentric values* – i.e., their apparent values at the actual place and time of observation – be this on the surface of the Earth, or aboard a spacecraft in space). If primes denote such topocentric values of the respective coordinates at a distance ϱ from the centre of the Earth and at a geocentric latitude φ, with Θ denoting the sidereal time of observation; while r and r' stand for the geocentric and topocentric distance of the Moon's centre, the exact equations relating the (instantaneous) topocentric right-ascension $\alpha'_{\langle}$ and declination $\delta'_{\langle}$ with their geocentric values (as tabulated in any standard ephemeris) are of the form

$$
\begin{aligned}
r'\cos\delta'_{\langle}\cos\alpha'_{\langle}&=r\cos\delta_{\langle}\cos\alpha_{\langle}-\varrho\cos\varphi\cos\Theta,\\
r'\cos\delta'_{\langle}\sin\alpha'_{\langle}&=r\cos\delta_{\langle}\sin\alpha_{\langle}-\varrho\cos\varphi\sin\Theta,\\
r'\sin\delta'_{\langle}&=r\sin\delta_{\langle}\qquad-\varrho\sin\varphi,
\end{aligned}\tag{2.13}
$$

which can be solved to yield the differences $\alpha'_{\langle}-\alpha_{\langle}$ and $\delta'_{\langle}-\delta_{\langle}$ from

$$\tan(\alpha'_{\langle}-\alpha)=\frac{\varrho\cos\varphi\sin\pi_{\langle}\sec\delta_{\langle}\sin(\alpha_{\langle}-\Theta)}{r-\varrho\cos\varphi\sin\pi_{\langle}\sec\delta_{\langle}\cos(\alpha_{\langle}-\Theta)}\tag{2.14}$$

and

$$\tan(\delta'_{\langle}-\delta_{\langle})=\frac{\varrho\sin\varphi\sin\pi_{\langle}\csc\Gamma\sin(\Gamma-\delta_{\langle})}{\varrho\sin\varphi\sin\pi_{\langle}\csc\Gamma\cos(\Gamma-\delta_{\langle})-r},\tag{2.15}$$

where, in the latter, the auxiliary angle Γ is defined by the equation

$$\tan\Gamma=\tan\varphi\cos\tfrac{1}{2}(\alpha'_{\langle}-\alpha_{\langle})\sec\tfrac{1}{2}(2\Theta-\alpha'_{\langle}-\alpha_{\langle});\tag{2.16}$$

and where the geocentric radius-vector r can be expressed in terms of the lunar equatorial horizontal parallax $\pi_{\langle}$ by means of the relation

$$r=\varrho\csc\pi_{\langle}.\tag{2.17}$$

Moreover, once the angles defined by the foregoing equations have been evaluated, the ratio of the topocentric and geocentric distance of the Moon results from

$$\frac{r'}{r} \frac{\sin(\delta_{\text{《}}-\Gamma)}{\sin(\delta'_{\text{《}}-\Gamma)};$$ (2.18)

so that the *augmentation* of the apparent semi-diameter s' of the lunar disc, as seen by the observer on the surface of the Earth, or in space, over its geocentric semi-diameter s is given by the equation

$$s'/s = r'/r.$$ (2.19)

As the observer on Earth can never be at a greater distance from the Moon than r if our satellite is to be visible above the horizon, it follows that $s' > s$.

All equations given so far in this section are exact, and should permit us to evaluate b, l and b', l' and the ratios s'/s or r'/r to the same precision to which the underlying data are known. However, for observations carried out on the Earth it is often sufficient to approximate the exact expression (2.12) for b by

$$b = -\beta_{\text{《}} - B + \tan^2 \tfrac{1}{2} I \sin 2(B+\beta_{\text{《}}) + \cdots,$$ (2.20)

where we have abbreviated

$$\tan B = \tan I \sin(\lambda_{\text{《}} - \Omega);$$ (2.21)

or, correctly to the squares of the small angle $I = 1°32'$, the geocentric optical libration in latitude can be obtained from

$$b + \beta_{\text{《}} = -I \sin(\lambda_{\text{《}} - \Omega) + \tfrac{1}{4} I^2 \sin 2\beta_{\text{《}} + \cdots,$$ (2.22)

while, similarly, the optical libration in longitude

$$\begin{aligned} l &= \lambda_{\text{《}} - l_{\text{《}} + bI \cos(\lambda_{\text{《}} - \Omega) + \tfrac{1}{4} I^2 \sin 2(\lambda_{\text{《}} - \Omega) + \cdots, \\ &= \lambda_{\text{《}} - l_{\text{《}} - \beta_{\text{《}} I \cos(\lambda_{\text{《}} - \Omega) - \tfrac{1}{4} I^2 \sin 2(\lambda_{\text{《}} - \Omega). \end{aligned}$$ (2.23)

The corrections necessary to reduce the geocentric values of b and l to the topocentric (primed) ones can be evaluated by taking advantage of the fact that, if highest accuracy is not required, the solutions (2.14) and (2.15) of Equations (2.13) can be approximated by the expressions

$$\alpha'_{\text{《}} - \alpha_{\text{《}} = -(\varrho/r) \cos\varphi \sin\pi_{\text{《}} \sec\delta_{\text{《}} \sin\Theta + \cdots,$$ (2.24)

$$\delta'_{\text{《}} - \delta_{\text{《}} = -(\varrho/r) \sin\pi_{\text{《}}(\sin\varphi \cos\delta_{\text{《}} - \cos\varphi \sin\delta_{\text{《}} \cos\Theta) + \cdots;$$ (2.25)

and, similarly, (2.19) can be approximated by

$$s' = s\{1 + (\varrho/r) \sin\pi_{\text{《}} \cos z\} \approx s\{1 + 0.0166 \cos z\},$$ (2.26)

where z denotes the zenith distance of the Moon at the time and place of observation and, as such, can be evaluated from known data with the aid of the formula

$$\cos z = \sin\varphi \sin\delta'_{\text{《}} + \cos\varphi \cos\delta'_{\text{《}} \cos(\alpha'_{\text{《}} - \Theta).$$ (2.27)

The topocentric values b' and l' then follow from the equations

$$\sin b' = \cos \sigma \sin b + \sin \sigma \cos b \cos(Q - C) \tag{2.28}$$

and

$$\sin(l' - l) = -\sin \sigma \sin(Q - C) \sec b', \tag{2.29}$$

where σ denotes the selenocentric angle between the observer and the centre of the Earth; and, as such, is given by the equation

$$\tan \sigma = \frac{\sin \gamma}{\csc \pi_\mathrm{c} - \cos \gamma}, \tag{2.30}$$

where γ is the geocentric angle between the observer and the Moon (i.e., geocentric 'zenith distance' of the Moon, obtainable from Equation (2.27) in which geocentric coordinates α_c, δ_c are used in place of topocentric ones); Q, the azimuth of the observer measured at the sublunar point; and C the position angle of the Moon's central meridian (measured positively eastward from the north). The latter two angles are, in turn, defined by the equations

$$\sin Q = \cos \varphi \csc \gamma \sin(\alpha_\mathrm{c} - \Theta),$$
$$\cos Q = (\sin \varphi - \cos \gamma \sin \delta_\mathrm{c}) \sec \delta_\mathrm{c} \csc \gamma; \tag{2.31}$$

and

$$\sin C = -\sin j \sec b \cos(\alpha_\mathrm{c} - \Omega') =$$
$$= \sin j \sec \delta_\mathrm{c} \cos(l + l' - \Omega + \varDelta). \tag{2.32}$$

Since, however, for in observer on the Earth $\pi_\mathrm{c} < 1°$ and, therefore, very approximately

$$\sigma = \pi_\mathrm{c} \sin \gamma, \tag{2.33}$$

Equations (2.28) and (2.29) can be readily approximated (cf.,e.g., Atkinson, 1951) by

$$b' - b = \pi_\mathrm{c} \sin \gamma \cos(Q - C) \tag{2.34}$$

and

$$l' - l = \pi_\mathrm{c} \sin \gamma \sin(Q - C) \sec b'. \tag{2.35}$$

The reader will doubtless appreciate now the fact that – even in their simplest form as given above – the corrections necessary to convert the geocentric librations of the Moon into topocentric librations are not exactly simple. While the geocentric values b, l of the Moon's optical libration, as defined by Equations (2.1) or (2.5), can usually be taken with sufficient precision from the existing ephemerides, their reductions to the topocentric values appropriate for a particular place and time of observation (or the converse operation) must be performed by the observer himself, with the aid of such formulae as given above. Fortunately, for an observer situated on the Earth (though *not* in space !) the absolute amounts of the differences $b' - b$ or $l' - l$ are small (always less than a degree) and need not, therefore, be usually computed to too many significant places.

The optical librations of the Moon represent phenomena of some mangitude, and were detected early in the history of lunar studies. The first one to have noticed the periodic displacement of lunar spots alternately toward the eastern and western limb appears to have been Galileo Galilei, who in his *Dialogues on the Two Great World Systems* had Salviati voice two possible causes: namely, the libration in latitude and diurnal libration – while the bulk of the phenomenon he observed was due to the libration in longitude. This libration was recognized as such by Riccioli and Hevelius between 1638–1641. In 1654, in a letter to Riccioli, Hevelius expressed first the opinion that the true cause of the observed libration in longitude is the non-uniformity of the motion of the Moon in its orbit (implicit, in fact, in Kepler's second law known since 1609). The correct explanation of all three optical librations of the Moon in terms of the characteristics of its orbit was advanced in the third volume of his *Principia* by Newton (who, incidentally, was the first to introduce the term 'optical libration' in this connection). Later in the same volume Newton, in discussing the problem of the figure of the Moon, mentioned also the possibility of its physical librations caused by the attraction of the Earth.

The announcement, in 1693, of Cassini's laws as stated at the commencement of this section followed the publication of the *Principia* by only six years; but as neither Giovanni Domenico Cassini himself, nor his son Jacques ever published the observations from which the three laws were derived, these did not exert their full impact on the contemporary scientific thought until they were later confirmed by Mayer (1748–1749) and Lalande (1763). For Newton, following Hevelius, the axis of rotation of the Moon was still perpendicular to its orbital plane (i.e., he took $I = 0$). Cassini was the first to establish that this inclination was finite, and adopted for it a value of $I = 2\frac{1}{2}°$, which Lalande reduced to $1°43'$.

By that time it had been realized that Cassini's laws, far from being accidental, must represent integrals of at least approximate equations of the problem of the Moon's motion about its centre of mass, in much the same way as Kepler's laws anticipated empirically certain closed integrals of the problem of two bodies; and – in the contemporary words of Tobias Mayer – 'anyone who could show the natural cause of this connection would be a happy and famous man, who would justly earn laurels for a new and great discovery'. The dynamical problem of the Moon's rotation arising from the existence of Cassini's laws was felt to constitute such a challenge that the Paris Academy of Sciences offered a special prize for its solution.

Both Euler and d'Alembert responded (rather unsuccessfully) to this challenge; but the laurels of which Mayer spoke were really carried away by Lagrange who won the Academy's prize in 1764; and who in two classical papers (the first submitted to the Paris Academy in 1768; the second – much more important – to the Berlin Academy in 1780) presented so complete a solution to the problem that subsequent investigations of Laplace (1798) or Poisson only filled in the details.

In the more recent past much more complete solutions of this problem have been constructed by a number of investigators – among whom two in particular should be remembered: namely, Friedrich Hayn (1863–1928) and Thaddeus Banachiewicz (1882–1954); and of the contemporary contributors to this subject we should mention Banachiewicz's pupil and successor Karol Koziel (1910–   ), Sh. T. Habibullin (1916–   ), D. H. Eckhardt (1932–   ), William M. Kaula (1926–   ) and M. D. Moutsoulas (1936–   ). In the light of all their work, how accurate are Cassini's laws, and what is the magnitude of the deviations by which Nature departs from them? The answer to this question, which exercised the minds of astronomers through-out the 18th and 19th centuries, is stored in the solution of the well-known differential equations of motion of a solid body about its centre of gravity, as formulated by Euler in the form

$$A\dot{\omega}_x - (B-C)\,\omega_y\omega_z = F_x, \tag{2.36}$$

$$B\dot{\omega}_y - (C-A)\,\omega_z\omega_x = F_y, \tag{2.37}$$

$$C\dot{\omega}_z - (A-B)\,\omega_x\omega_y = F_z, \tag{2.38}$$

where

$$\omega_x = -\dot{\psi}\sin\theta\sin\phi - \dot{\theta}\cos\phi, \tag{2.39}$$

$$\omega_y = -\dot{\psi}\sin\theta\cos\phi + \dot{\theta}\sin\phi, \tag{2.40}$$

$$\omega_z = \quad \dot{\psi}\cos\theta \qquad\qquad +\dot{\phi}, \tag{2.41}$$

denote the angular velocities of rotation about a system of selenocentric rectangular coordinates, the axes of which coincide with the principal axes of inertia of the lunar globe, expressed in terms of the three Eulerian angles θ, ϕ, ψ of generalized rotation; while A, B, C denote the moments of inertia of the Moon about its principal axes; and lastly, $F_{x,y,z}$ stand for the components of external forces (in our case, the attrac-tion of the Earth and of the Sun) which act on our rotating configuration.

If Cassini's laws were exact, the integrals of motion satisfying Equations (2.36)–(2.38) should obviously be expressible as

$$\theta = I, \tag{2.42}$$

$$\psi = \Omega, \tag{2.43}$$

$$\phi = 180° + l - \Omega, \tag{2.44}$$

where l_{c} denotes, as before, the mean longitude of the Moon in its orbit. In order to investigate the extent to which the actual motion of the Moon may depart from these conditions, let us generalize the foregoing Equations (2.42)–(2.44) by setting

$$\theta = I + \varrho, \tag{2.45}$$

$$\psi = \Omega + \sigma, \tag{2.46}$$

$$\phi = 180° + l_{\text{c}} - \psi + \tau, \tag{2.47}$$

where ϱ, σ and τ represent the so-called *physical librations* of the Moon in latitude, node and longitude; and on insertion of (2.45)–(2.47) in (2.39)–(2.41) consider (2.36)–(2.38) as differential equations for ϱ, σ and τ.

A construction of the actual solutions of such equations is somewhat involved, and need not be reproduced in this place; the reader desirous to get acquainted with its details may consult Chapter 4 of Kopal (1969). The outcome discloses (cf. Eckhardt, 1965), however, the leading terms of the expansions for ϱ, σ and τ to be of the form

$$\varrho = -99\overset{\prime\prime}{.}0\,\cos g + \cdots, \tag{2.48}$$

$$I\sigma = -101\overset{\prime\prime}{.}3\,\sin g + \cdots, \tag{2.49}$$

$$\tau = \quad 91\overset{\prime\prime}{.}5\,\sin g' + \cdots, \tag{2.50}$$

where g denotes the mean anomaly of the Moon; and g', that of the Sun. Therefore – unlike the optical librations of our satellite which may attain several degrees of arc from our terrestrial vantage point – the selenocentric physical librations of the lunar globe do not, in general, exceed two minutes. The leading terms in ϱ and σ arise from the Moon's 'elliptic inequality'; while the leading term of τ is due to the 'annual equation'.

The smallness of the Moon's physical librations becomes even more manifest when we consider the visibility of their effects from the Earth. As the radius of the lunar globe is 221 times smaller than the mean distance separating us from our satellite, a selenocentric libration of $2'$ would be seen from the Earth as an angular displacement of $2'/221 = 0\overset{\prime\prime}{.}54$ at the centre of the apparent linar disc, and progressively less towards its limb. No wonder that motions so small had to await their discovery long after Newton and Lagrange predicted their existence! The fact that this proved possible in the 19th century testifies to the skill and patience of the observers like Ernst Hartwig (1851–1923) who dedicated his entire life to this exacting task.

References

Atkinson, R. d'E.: 1951, *Monthly Notices Roy. Astron. Soc.* **111**, 448.
Brown, E. W.: 1919, *Tables of the Motion of the Moon*, Yale Univ. Press, New Haven, Conn.
Eckhardt, D. H.: 1965, *Astron. J.* **70**, 466.
Kopal, Z.: 1969, *The Moon*, D. Reidel Publ. Co., Dordrecht, Holland, Chapter 4.
Koziel, K.: 1967a, *Icarus* **7**, 1–28.
Koziel, K.: 1967b, in *Measure of the Moon* (ed. by Z. Kopal and C. L. Goudas), D. Reidel Publ. Co., Dordrecht, Holland, pp. 3–11.
Lipsky, Yu. N.: 1965, *Sky Telesc.* **30**, 338.

CHAPTER 3

SELENOGRAPHIC COORDINATES

With all essential facts on the generalized rotation of the lunar globe on record in the preceding chapter, all prerequisites are now available for the adoption of a specific system of lunar coordinates.

The most appropriate system of such coordinates – in fact, the only system whose coordinates are rigidly fixed in the Moon, and do not change in time for any observer stationed on its surface – would be those whose axes are identified with the principal axes of inertia of the lunar globe. For a figure of the Moon described in simple geometrical terms – ellipsoidal, for instance – the axes of symmetry could be identified with the principal axes of the respective ellipsoid. As we shall explain in Chapter 4, however, the actual figure of the Moon (or, more accurately, its departures from a sphere – no matter how small these may be) are quite complicated, and not describable in terms of any rapidly convergent series of spherical harmonics. Under these conditions, the exact positions of the principal axes of inertia of the lunar globe could be determined only from astrometric observations made on the lunar surface; and not by those made from any external vantage point.

Such being the case, the best way for defining a coordinate system on the Moon is to base it on the position of the lunar axis of rotation, and on the equatorial plane which is normal to it. In such a system, the angular coordinates (i.e., the latitude and longitude) of any surface point will remain constant and independent of the time; but they will not be time-independent with respect to an inertial system – such as one referred to the ecliptic as one of its principal planes.

The reason is the fact that – for reasons explained already in the preceding chapter – the direction of the lunar axis of rotation is not fixed in space, but is subject to both precession and nutation which displaces constantly its direction with respect to the ecliptic. In more specific terms, the direction cosines of the true axis of rotation of the Moon at any time will evidently be given by the expressions

$$-\sin\phi\,\sin\theta, \qquad -\cos\phi\,\sin\theta, \qquad \cos\theta, \tag{3.1}$$

where the Eulerian angles θ and ϕ continue to be defined by Equations (2.39)–(2.41). Since, however, the latter angles contain the periodic time-dependent terms ϱ, σ and τ constituting the physical librations of the lunar globe, the position of the pole of rotation will clearly oscillate in space – in the course of each lunar 'day' – in accordance with Equations (2.48)–(2.50).

Let, in what follows, the *selenographic latitude* β be the angular distance between the respective point and the equator, measured in the meridional plane positively northward (i.e., towards the hemisphere containing Mare Imbrium); and the seleno-graphic longitude λ be measured in the plane of the equator positively eastward (i.e., towards Mare Crisium); the zero-point from which the longitude is the direction of the Moon at its ascending node at the time of the perigee (or apogee).

The axes of so-defined selenographic longitude and latitude are fixed neither with respect to the Moon (since the axis of rotation may migrate within the lunar globe – as it does in the Earth – with respect to its principal axes of inertia), nor with respect to the ecliptic. In actual fact, the difference between a lunar system of coordinates based on the principal axes of inertia and the instantaneous equator are so small that we can ignore it for practical purposes and regard the angular coordinates based on the instantaneous position of the lunar equator as the 'true' selenographic latitude and longitude.

The mean position of the lunar axis of rotation in the course of each month will continue to be specified by the direction cosines (3.1), in which we insert for the Eulerian angles θ and ϕ from (2.42)–(2.44) rather than from (2.45)–(2.47) – i.e., in which the physical librations ϱ, σ and τ specifying short-period time oscillations have been ignored. The selenographic coordinates λ_0, β_0 referred to the mean (Cassini) axes are termed in the literature as the 'mean' selenographic coordinates; and are sometimes (though not frequently) used in selenographic literature under this name.

The relation between the 'true' and 'mean' selenographic coordinates is provided by the equations

$$\lambda - \lambda_0 = \tau - \{ \varrho \cos(\lambda + l - \Omega) + I\sigma \sin(\lambda + l - \Omega) \} \tan\beta \qquad (3.2)$$

and

$$\beta - \beta_0 = \varrho \sin(\lambda + l - \Omega) - I\sigma \cos(\lambda + l - \Omega). \qquad (3.3)$$

Since (see Equations (2.48)–(2.50) of the preceding chapter) the selenocentric physical librations do not exceed, in general, two minutes of arc, the differences between 'mean' and 'true' selenographic coordinates may amount to a little more than one kilometre on the lunar surface (a distance which, viewed from the Earth, would corresond to an angle of $0\rlap{.}{''}54$).

The distinguishing feature of the 'mean' coordinates λ_0 and β_0 rests on the fact that, unlike λ and β, the mean coordinates are immune to short-periodic oscillations with respect to an external observer; but they do not remain fixed in space either. By virtue of the neglect of the physical librations ϱ, σ, τ, the 'mean' coordinates λ_0, β_0 do not 'nutate'; but they do 'precess' – on account of the fact that Ω in (2.1) or (2.5) recedes in synchronism with the motion of the ascending node of the lunar orbit; and completes one revolution in 248.83 sidereal months or 18.61 tropical years; but in the course of any one month their motions are much smaller.

With these definitions kept in mind, let us turn to the problem of determination of the selenographic longitude λ and latitude β of any arbitrary point $P(\lambda, \beta)$ on the Moon for which the rectangular coordinates x, y have been measured on (say) a photographic plate, taken at a time at which the topocentric librations l', b' of the

Moon (whether they pertain to any location on the Earth or in space) are known. In order to do so, let (cf. Figure 3.1)

$$\Pi = \tan^{-1}(x/y) \tag{3.4}$$

denote the position angle of any point $P(r, \lambda, \beta)$ on the apparent lunar disc (i.e., in the xy-plane), measured counter-clockwise from the North Pole; C'_1, the (topocentric)

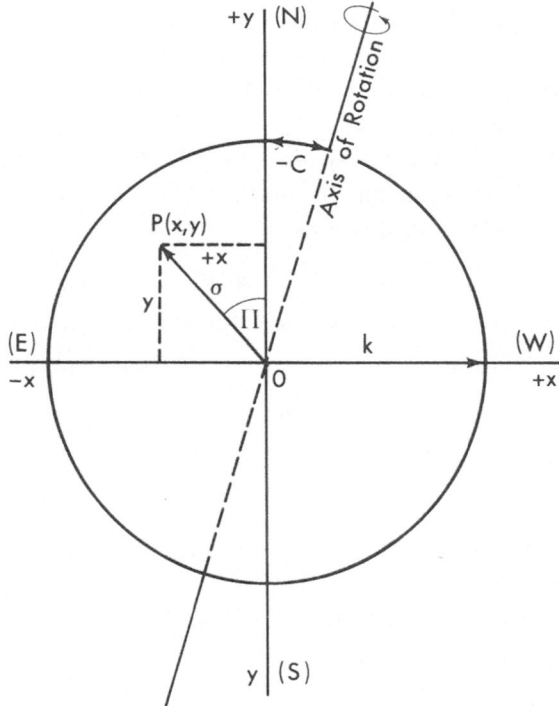

inclination of the Moon's axis of rotation (obtainable from its geocentric value as given by Equation (2.32)); and

$$\kappa = \sin^{-1}\{(r/k)\cos s'_i\}, \tag{3.5}$$

with s'_i standing for the angular semi-diameter of the apparent lunar disc at the time of observation (which, for Earth-based work, is given by Equation (2.19)).

With these definitions, the equations defining the selenographic coordinates λ_P and β_P of the point P are of the form

$$\sin\beta_P = \sin b' \cos\kappa + \cos b' \sin\kappa \cos(\Pi - C'),$$
$$\cos(\lambda_P - l')\cos\beta_P = \cos b' \cos\kappa - \sin b' \sin\kappa \cos(\Pi - C'), \tag{3.6}$$
$$\sin(\lambda_P - l')\cos\beta_P = \qquad + \qquad \sin\kappa \sin(\Pi - C').$$

Moreover, if we set

$$\tan \kappa \, \cos(\Pi - C') = \tan Q, \tag{3.7}$$

the latter two Equations of (3.6) can be solved for λ_P to yield

$$\tan(\lambda_P - l') = \tan(\Pi - C') \sec(Q + b') \sin Q; \tag{3.8}$$

while β_P follows from the first one of (3.6).

With the topocentric librations l', b' as given in Chapter 2, the λ_P, β_P's so obtained should represent the 'mean' selenographic coordinates, referred to the Cassini equator; whereas if l', b' as used incorporate also the effects of physical librations, 'true' selenographic coordinates result; otherwise the latter can be obtained from the former by a recourse to (3.2) and (3.3).

Conversely, should the values of λ_P, β_P as well as the libration constants l' and b' be given, and the task be to evaluate the position of the respective point on the lunar face in the xy-coordinates, the equations

$$\cos \kappa = \sin b' \, \sin \beta_P + \cos b' \, \cos \beta_P \, \cos(\lambda_P - l'),$$
$$\cos(\Pi - C') \sin \kappa = \cos b' \, \sin \beta_P - \sin b' \, \cos \beta_P \, \cos(\lambda_P - l'), \tag{3.9}$$
$$\sin(\Pi - C') \sin \kappa = \qquad\qquad + \qquad \cos \beta_P \, \sin(\lambda_P - l'),$$

can be used to evaluate κ and Π, from which

$$x = k \sec s' \, \sin \kappa \, \sin \Pi \tag{3.10}$$

and

$$y = k \sec s' \, \sin \kappa \, \cos \Pi. \tag{3.11}$$

The foregoing systems of Equations (3.6) or (3.9) can be re-stated also in the following more economic way. Let

$$\xi = \cos \beta_P \, \sin \lambda_P,$$
$$\eta = \sin \beta_P, \tag{3.12}$$
$$\zeta = \cos \beta_P \, \cos \lambda_P,$$

be the direction cosines of the radius from the Moon's center to the point P. If so, Equations (3.6) can evidently be rewritten as

$$\xi = \quad x \cos l' - y \, \sin b' \, \sin l' + z \, \cos b' \, \sin l',$$
$$\eta = \qquad\qquad + y \cos b' \qquad + z \, \sin b' \tag{3.13}$$
$$\zeta = -x \, \sin l' - y \, \sin b' \, \cos l' + z \, \cos b' \, \cos l',$$

where, for a spherical Moon,

$$x^2 + y^2 + z^2 = 1. \tag{3.14}$$

Consequently, by their inversion it follows that

$$
\begin{aligned}
x &= \quad \xi \cos l' \qquad\qquad\qquad - \zeta \, \sin l', \\
y &= -\xi \, \sin l' \, \sin b' + \eta \, \cos b' - \zeta \, \cos l' \, \sin b', \\
z &= \quad \xi \, \sin l' \, \cos b' + \eta \, \sin b' + \zeta \, \cos l' \, \cos b'.
\end{aligned}
\tag{3.15}
$$

The direction cosines ξ, η, ζ may evidently be regarded as the rectangular coordinates of a point P at unit distance from the Moon's center; the ζ-axis being oriented toward the Earth along the lunar 'first radius' (defined as the intersection of the equator and prime meridian), and the η-axis coinciding with the Moon's axis of rotation. Their values are often used to describe the position of a point on the lunar surface in place of the angular coordinates λ_P and β_P.

If, for some purpose, it were required to obtain the selenocentric right-ascension α_P and declination δ_P on the lunar surface, of known selenographic coordinates λ_P, β_P, this can be done with the aid of the equations

$$
\begin{aligned}
\cos \delta_P \cos(\alpha_P - \Omega') &= \cos \beta_P \cos \lambda_P, \\
\cos \delta_P \sin(\alpha_P - \Omega') &= \cos \beta_P \sin \lambda_P \cos i - \sin \beta_P \sin i, \\
\sin \delta_P \quad &= \cos \beta_P \sin \lambda_P \sin i + \sin \beta_P \cos i,
\end{aligned}
\tag{3.16}
$$

where i denotes the angle of inclination between the lunar and terrestrial equators; and Ω', the longitude of the ascending node of the lunar equator, as given by Equation (3.6). The selenocentric ecliptical coordinates L_P, B_P would follow, similarly, from the transformation equations

$$
\begin{aligned}
\cos B_P \cos(\Omega - L_P) &= -\cos \beta_P \cos(\Omega - \lambda_P - l_\odot) \\
\cos B_P \sin(\Omega - L_P) &= -\cos \beta_P \sin(\Omega - \lambda_P - l_\odot) \cos l + \sin \beta_P \sin l, \\
\sin B_P \qquad &+ = \cos \beta_P \sin(\Omega - \lambda_P - l_\odot) \sin l + \sin \beta_P \cos l,
\end{aligned}
\tag{3.17}
$$

where Ω denotes, as before, the mean longitude of the ascending node of lunar orbit on the ecliptic; $l_\odot$, the mean longitude of the Sun; and I, the inclination of the lunar equator to the ecliptic. The two systems L_P, B_P and λ_P, β_P differ only in so far as the latitude B_P is measured from a plane passing through the center of the Moon and parallel with the ecliptic, while β_P is measured from the lunar equator; and L_P is measured from the vernal equinox, while λ_P, from the lunar prime meridian.

The last remaining task which remains to be settled before our lunar coordinates become uniquely determined is to specify the zero-point from which the longitudes are to be measured – i.e., the point at which the 'first radius' intersects the lunar surface. This can be done only from the observed lunar librations; and efforts to do so go back to Tobias Mayer.

As this zero-point is not marked by any particular feature, the customary procedure has been to measure with respect to it the position of some unambiguous and well-defined feature in its proximity. Tobias Mayer – the father of positional selenography – chose for this purpose the central peak of the crater Manilius (a practice in which he was followed by Bouvard and Pécuchet); but since the days of Bessel, the feature adopted for this purpose (on account of the regularity of its form) was a small crater Mösting A (see Figure 3.2), for the position of which Koziel (1967) established, by an extensive rediscussion of all heliocentric observations made between 1877 and 1915, the true selenographic coordinates to be

$$\lambda = 5° \ 9'53'' \pm 5'' \ (\text{m.e.}) \ \text{W},$$
$$\beta = 3°10'41'' \pm 4'' \ (\text{m.e.}) \ \text{S}.$$

(3.18)

It may be added that an uncertainty of $\pm 4''$ to $5''$ in the respective lunar coordinates corresponds, on the lunar surface, to a linear uncertainty of ± 35 to 40 m (about the same as is inherent in the present uncertainty of the location of the lunar poles).

Fig. 3.2.   Position of Mösting A.

The foregoing values of λ and β for Mösting A, together with Koziel's value of $I = 5524'' \pm 7''$ for the mean value of the inclination of the Moon's axis of rotation to the ecliptic, have remained at the basis of the 'true' selenographic system of co-ordinates up to the present time. Quite recently, Wollenhaupt *et al.* (1972) deduced a new value of $I = 5659''$ (by $135''$ larger than Koziel's) from an analysis of theodolite measurements of lunar features from the command module of the Apollo 15 mission in August 1971. Since Koziel's value refers to a mean equinox close to 1900, while the new one was obtained for 1971.6, their comparison suggests a secular increase in I by $135''$ in yr, or about $2''$ per annum; but new observations from additional Apollo missions will be needed to place such a conjecture on firmer footing.

Based on the measured positions of Mösting A, additional control points on the lunar surface have been established by different investigators. Thus Julius Franz (1847–1913) used – besides Mösting A – the positions of the craters Aristarchus, Byrgius A, Fabricius K, Gassendi, Macrobius A, Nicollet A, Proclus, and Sharp A as a system of nine fundamental reference points on the lunar surface; while Friedrich Hayn (1863–1928) used only five (Egede A, Kepler A, Messier A, Mösting A, and Tycho). The positions of some of these reference points were more recently remeasured by Schrutka-Rechtenstamm (1956, 1958).

The details just mentioned constitute a group of lunar triangulation points of first order, based (very largely) on heliometric measurements. With their aid, the positions of a much larger number of secondary points were determined, partly visually (Franz, 1901), but mainly photographically (Saunder, 1905, 1911; König and others). Franz's list contains the positions of a total of 1446 secondary reference points on the surface of the Moon. Saunder measured 2885 of them from negatives secured by Loewy and Puiseux with the equatoreal coudé at Paris (Saunder, 1905), and by Ritchey with the 40-in. Yerkes refractor (Saunder, 1911); while König's results, while more numerous, were based on measurements of plates taken with a much smaller instrument (8-in. refractor of 343 cm focal length). Saunder's measurements were recently used to furnish, by graphical interpolation, the coordinate net of the *Orthographic Atlas of the Moon* (Arthur *et al.*, 1961); and the errors inherent in such a process were subsequently discussed by Hawkins (1963). For the latest list of the determinations of the selenographic coordinates of 700 lunar points cf. Moutsoulas (1972).

In the past few years, more accurate lunar control systems than anything previously obtainable from the distance of the Earth have been established on the basis of space-born Orbiter and Apollo photography at the Aeronautical Chart and Information Center (ACIC) of the U.S. Air Force, by a team of investigators (see Figures 3.3 and 3.4) led by Robert W. Carder and Lawrence Schimerman.

Development of selenographic positions from photographs acquired by spacecraft operating in lunar proximity is similar to earthbased telescopic solutions in that both rely upon steroscopic principals.

Spacecraft coordinate determinations depend on photogrammetric intersection with the stereoscopic base being provided by change in position of the spacecraft as contrasted to reliance on lunar libration in Earth-based work. Spacecraft based photogrammetric solutions have inherently greater precision by virtue of the superior geometry and larger scale photography involved.

The absolute basis for selenographic coordinates determined from spacecraft photography by photogrammetric triangulation is dependent on knowledge of exposure station positions and camera orientations. Conventional aerial triangulation as practiced in Earth surveys uses photoidentifiable geodetic positions for datum definition. However, the general absence of such precise control points in the lunar case makes this approach presently inapplicable. The required values for exposure station position with respect to the center of lunar mass are provided by a spacecraft ephemeris developed through integration of earthbased radar observations of the

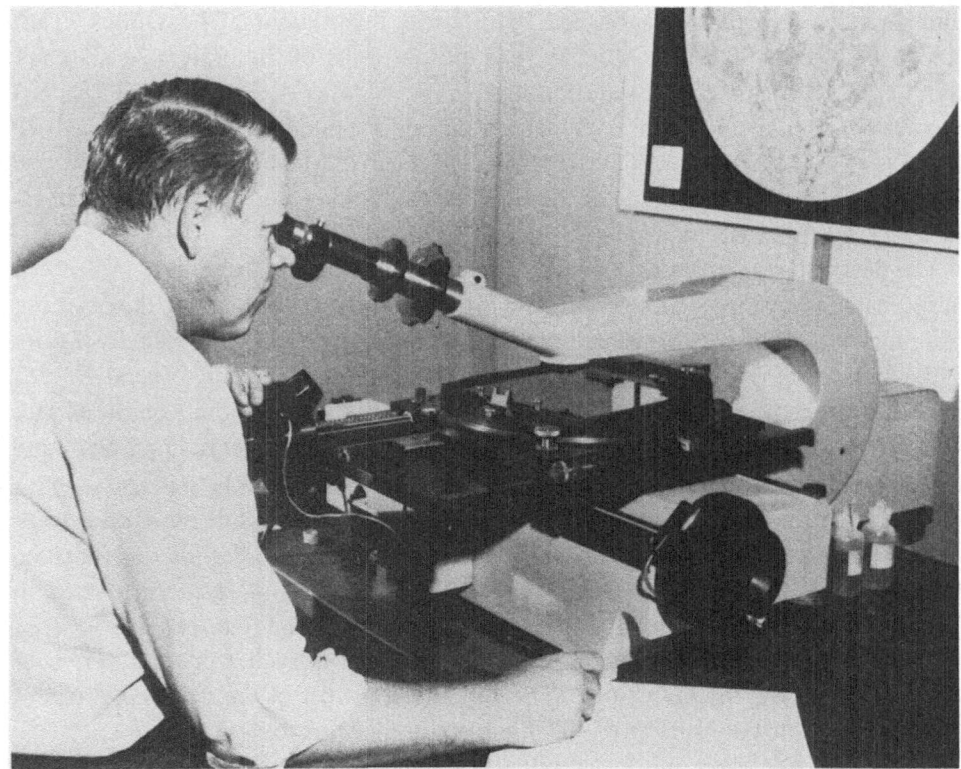

Fig. 3.3.   Don L. Meyer of ACIC making photogrammetric measurements of the Moon with
a Mann comparator.

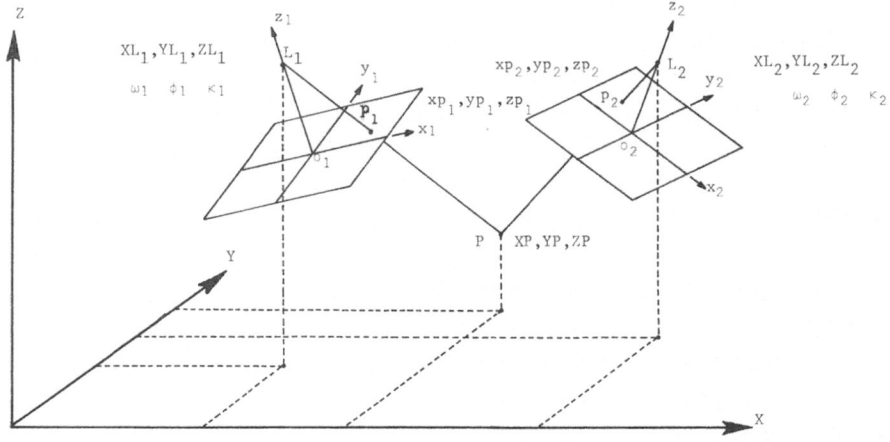

Fig. 3.4.   Geometry of lunar photogrammetry.

spacecraft. Spacecraft ephemeris computation employs a planetary ephemeris (defining Earth-Moon positional relationship with respect to time) and lunar libration and gravity models. Required camera orientation angles are determined with respect to stellar reference through companion stellar cameras, star trackers or spacecraft inertial measurement units. Figure 3.4 portrays the geometry of determining lunar coordinates by photogrammetric intersection in which:

L_1 and L_2 are the camera stations from which two overlapping photographs of the lunar surface were taken.

XL_1, YL_1, ZL_1 and XL_2, YL_2, ZL_2 are the rectangular coordinates of the respective camera stations in a selenocentric system defined by reference of exposure time to developed spacecraft ephemeris.

$\omega_1, \phi_1, \kappa_1$ and $\omega_2, \phi_2, \kappa_2$ are the camera orientation angles with respect to the selenocentric rectangular coordinate axes. This data is defined with respect to celestial reference by photographic or other stellar sensors having a calibrated relationship with the camera used to photograph the lunar surface.

xp_1, yp_1, and xp_2, yp_2, are the measured values of an image of the same lunar feature, (p_1, p_2) in each photograph's coordinate system.

$\overline{L_1o_1} = \overline{L_2o_2} = f =$ focal length of the taking camera known from camera calibration. The direction cosines of line L_1p_1 are

$$a_1 = \frac{xp_1}{(xp_1^2 + yp_1^2 + f^2)^{1/2}},$$

$$b_1 = \frac{yp_1}{(xp_1^2 + yp_1^2 + f^2)^{1/2}},$$

$$c_1 = \frac{zp_1}{(xp_1^2 + yp_1^2 + f^2)^{1/2}}.$$

Direction cosines of line $L_2p_2 = a_2, b_2, c_2$ are similarly derived.

Transformation of the direction cosines from the photographic to selenocentric coordinate systems through camera orientation angles $\omega_1, \phi_1, \kappa_1$ and $\omega_2, \phi_2, \kappa_2$, yields direction cosines A_1, B_1, C_1 for line L_1P and A_2, B_2, C_2 for line L_2P.

Coordinates of lunar feature points X_p, Y_p, Z_p may be obtained from simultaneous solution of pairs of equations from (1) and (2) as

(1) $$\frac{X - XL_1}{A_1} = \frac{Y - YL_1}{B_1} = \frac{Z - ZL_1}{C_1},$$

(2) $$\frac{X - XL_2}{A_2} = \frac{Y - YL_2}{B_2} = \frac{Z - ZL_2}{C_2}.$$

In practice, large numbers of overlapping photographs may be triangulated and varying constraints applied to photo measurements, exposure station positions and camera orientations depending upon the quality and variety of available data.

Early (1966–1970) photogrammetric triangulation from spacecraft photography was conducted at ACIC by Schimerman, Ross and Klute; at TOPOCOM by Rose, Shull and Schenk and at NASA/MSC by Norman, Hill and Hancock. Work was concentrated on development of lunar coordinates for potential lunar landing sites and areas of particular scientific interest, using Lunar Orbiter Mission photographs. Developed lunar control data were used in support of Apollo mission planning and execution as well as large scale lunar mapping. Triangulation from Lunar Orbiter photographs and data entailed reconstruction of photographic geometry, due to the segmented form in which Lunar Orbiter photographs were transmitted to Earth. Camera orientation was provided through a star tracker carried on board the unmanned Lunar Orbiter spacecraft.

In 1969, Ruffin and Schimerman at ACIC used Lunar Orbiter photographs to establish a Positional Reference system on the lunar far side to support small scale mapping. Perspective projections of a selenographic graticule were computed based on camera focal length, orientation, and position, and fitted to precision reassemblies of 24 selected photographs. The graphic best overall fit of projections to photographs and to images of Earth-based control points from the ACIC Selenodetic System of 1965 (Meyer and Ruffin) was the basis for this extension of coordinates to the lunar far side. The developed far-side positions are reflected in the 1:5 000 000 and 1:2 750 000 lunar map series prepared by ACIC for NASA. This approximate solution did not provide for determination of lunar elevations and has an evaluated accuracy of 5 to 20 km.

During 1969–1972, the previously named photogrammetric groups at ACIC, TOPOCOM and NASA performed triangulation with strips of Hasselblad photographs acquired on Apollo missions 8, 10, 12 and 14 to develop lunar coordinates for following Apollo landing sites and approach areas. Data available for these works was severely limited in that adequate camera calibration information was unavailable for the Hasselblad cameras and no precise time of exposure was recorded to allow direct determination of camera station positions from the spacecraft ephemeris. Landmark tracking points based on astronaut sextant observations developed by Wollenhaupt and Ransford at NASA/MSC, were used to contrain these triangulations. Camera orientations were based on spacecraft Inertial Measurement Unit data which was periodically updated by astronaut stellar observations.

The real potential for selenodetic improvement through photogrammatric triangulation, lies in the photographs and data acquired from Apollo Missions 15, 16 and 17. These missions provide coverage of approximately 20% of the lunar surface between 30 deg N and S lat. The Apollo Metric Camera System's recording of precise time of exposure (1 ms) of companion frame terrain and stellar photographs provide an accurate relation to spacecraft ephemeris to establish exposure station position as well as the means of determining camera orientation with respect to celestial reference. In addition, the laser altimeter component of the Metric System simultaneously records exposure station to lunar surface distance.

The photogrammetric triangulation of Apollo Mission 15 photography and data conducted by Schimerman, Helmering, Cannell and Hassle was completed in April

1973 at the Defense Mapping Agency Aerospace Center (formerly ACIC). The resulting coordinate system will serve as the control basis for a NASA lunar map series at 1:250000 scale, to be compiled in 1973–1975. Results obtained from the Apollo 15 photogrammetric reduction reflect an ability to project exposure station positions (defined by the spacecraft ephemeris) to the lunar surface with an accuracy of 20 to 40 m. This range is largely a function of the photographic resolution as affected by variance in Sun elevation at the time of photographic exposure. The stellar photographic reduction provides a good contribution to coordinate determination with identification and mensuration of 20 to 40 stellar images resulted in recovery of camera orientation to within 20".

The absolute accuracy of the Apollo 15 Control System is dependent on the validity of spacecraft ephemeral data inputs and validation of the selenodetic accuracy of this control work is a subject of further study.

References

Franz, J.: 1901, *Mitt. Sternwarte Breslau* **1**, 1–48.

Habibullin, Sh. T.: 1971, *Moon* **3**, 231–238.

Hawkins, G. S.: 1963, *Astron. Contr. Boston Univ., Ser. II*, No. 36.

Moutsoulas, M. D.: 1972, *Moon* **5**, 302–331.

Saunder, S. A.: 1905, *Monthly Notices Roy. Astron. Soc.* **65**, 458.

Saunder, S. A.: 1911, *Mem. Roy. Astron. Soc.* **60**, 1.

Schrutka-Rechtenstamm, G. von: 1956, *Sitzungsber. Österr. Akad. Wiss. (Math.-Naturwiss. Klasse)* **165**, 97–126.

Schrutka-Rechtenstamm, G. von: 1958, *Sitzungsber. Österr. Akad. Wiss. (Math.-Naturwiss. Klasse)* **167**, 71–106.

Wollenhaupt, W., Osburn, R. K., and Ransford, G. A.: 1972, *Moon* **5**, 149–157.

CHAPTER 4

SHAPE OF THE MOON

In the preceding chapter we defined the angular coordinates λ and β which specify the direction of the radius-vector from the centre of the Moon to the respective point. In order to specify completely the position of that point in space, the third – radial – coordinate is required which specifies the distance of that point from the Moon's centre (or, which is equivalent, its elevation h above the mean spherical surface of the Moon).

The most readily available information concerning the form of our satellite is provided to us by the shape of its *marginal* zone – i.e., of the region which appears in projection on the celestial sphere as the *limb* of the apparent lunar disc. The phenomena of lunar librations, described in the preceding chapter, will bring from time to time almost 17.7% of the entire lunar surface to the limb of its apparent disc in the sky; and the shape of this limb can be measured by the usual astrometric methods.

The best conditions for measuring the exact shape of the entire lunar limb present themselves during annual eclipses of the Sun by the Moon. In the course of such eclipses, the entire lunar circumference can be measured with reference to the adjacent solar limb (which is known to be circular within $0''.1$); and some recent observations of this type (cf. Carson *et al.*, 1967; Davidson *et al.*, 1967) have demonstrated the power of this method.

Under conditions obtaining at night-time the measurable arc of the limb will extend, in general, over only one-half of the entire circumference; for the other half of the 'meridian of illumination' is bound to be vitiated by the phase effect. This is true even at 'full' Moon unless both libration angles happen to vanish at the same time; and when this occurs the Moon undergoes eclipse in the shadow of the Earth.

On the basis of visual observations or lunar photography made at night-time, extensive maps of the marginal zones of the Moon which can appear on the limb have been prepared by Hayn (1907), Weimer (1952), Nefediev (1957), Gorynia and Drofa (1962) and Watts (1963). A harmonic expansion of the apparent form of the lunar meridian perpendicular to the line of sight at the time of zero libration, as defined by the ensemble of all observational data from Hayn through Watts (and also the more recent eclipse data) led, at the hands of Goudas (1965b) to an expansion of the apparent radius of the Moon of the form

$$r'' = 914''.61 \pm 0''.01$$
$$- (0''.25 \pm 0''.05) \sin \beta - (0''.07 \pm 0''.08) \cos \beta$$
$$+ (0''.46 \pm 0''.09) \sin 2\beta - (0''.24 \pm 0''.08) \cos 2\beta$$

$$+ (0\rlap{.}''29 \pm 0\rlap{.}''05) \sin 3\beta - (0\rlap{.}''07 \pm 0\rlap{.}''04) \cos 3\beta$$
$$- (0\rlap{.}''19 \pm 0\rlap{.}''08) \sin 4\beta + (0\rlap{.}''16 \pm 0\rlap{.}''05) \cos 4\beta + \cdots, \tag{4.1}$$

at the mean distance of the Moon, rendering *the limb of the Moon a deformed ellipse, elongated* (rather than compressed) *along an axis inclined by about* $35° \pm 2°$ *to the lunar axis of rotation.* All previous investigators of this subject (Yakovkin, 1952; Dommanget, 1962; Potter, 1962; and Watts, 1963) are in essential agreement on this fact. Yakovkin estimated independently this inclination to be $23°$; Watts, to $35°$; Potter, to $34°$; while Davidson *et al.* to $37°$; and all agree that this inclination lies in the same quadrant (i.e., from SW to NE). The orientation of the semi-major axis of symmetry deviates, therefore, markedly from the lunar axis of rotation; and the angular difference between the two axes exceeds one second of arc.

At the mean distance of the Moon from the Earth of 384 400 km, the measured apparent radius of our satellite corresponds to an absolute mean radius of 1738 km; though local deviations from it amounting to ± 4 km have been established by the observations. Within quantities of the order of 0.2% of its mean radius the Moon can, therefore, be regarded as a sphere; through already the limb measurements have disclosed that, within quantities of the order of ± 4 km, the deviations of the Moon's actual surface from a sphere are likely to be quite complicated.

In order to ascertain the magnitude of such deviations over the entire face of the Moon, we must resort to other methods than those mentioned so far, whose applicability is not restricted to the limb. Two such methods lend themselves, in principle, to this task. These are:

(1) *Stereoscopic techniques* – based on the observations of the angular displacements of the individual points P on the lunar surface, with respect to the limb, or to an absolute frame of reference as represented (for instance) by the background stars, in the course of lunar librations; and

(2) *Ranging techniques* – i.e., determination of the absolute distance of P from an observing station at known distance from the centre of the Moon (be it located on the Earth, or aboard the spacecraft operating in closer proximity of the lunar surface).

Until the advent of spacecraft in the past few years, the method (1) alone was available for investigations of the shape of the lunar surface, by observations made from the distance of the Earth. Since, moreover, this method still provided most of what we know of the shape of the Moon as a whole, in what follows an account of its underlying geometry will be given – all the more so as the method is equally applicable (by a simple transfer of the observing station from the Earth to any other point in space – such as that occupied by a spacecraft at a particular time) to a stereoscopic analysis of photographs taken from closer proximity to the lunar surface than that separating the Earth and the Moon.

Let P in Figure 4.1 be the point whose absolute coordinates are to be determined, S the intersection of the line of sight of the terrestrial observer at E with a mean sphere, and D the projection of P (not corrected for finite distance) on the disc LPQ. The elevation PS of the point P above or below the mean sphere will be denoted again by h. Let QU be the projection of the axis of rotation of the Moon on the visible disc at that time. We shall now define a 'librated' rectangular frame of reference $Mxya$ whose origin rests at the point M, whose Mxy plane coincides with the plane of the lunar disc, and whose Mx-axis makes an angle $\pi/2 - \theta$ with the projection of the axis of rotation NN' on the disc. This angle can be determined with the aid of the topo-

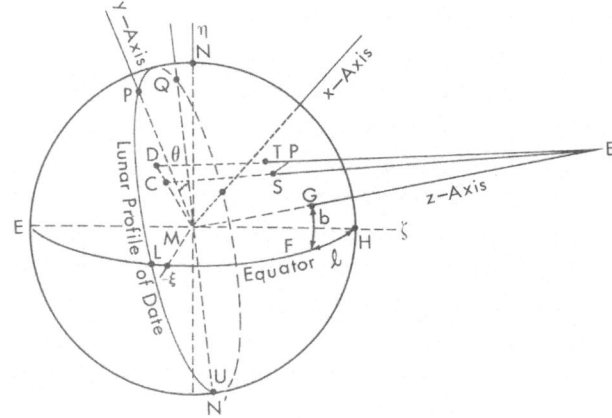

Fig. 4.1.   Application of the principle of stereoscopy in a general case.

centric libration angles $\sphericalangle HF = l$ and $\sphericalangle FG = b$, after the direction of the Mx-axis is arbitrarily fixed on the photographic plate. Finally, the Mz-axis is arbitrarily fixed on the photographic plate, while the Mz-axis coincides with the direction ME and is positive towards the observer E.

If now, x, y, z are the coordinates of the point T and x_1, y_1, z_1 the coordinates of point S fn the $Mxyz$ frame of reference, then the coordinates of the point P will be

$$x_1 + x_1/r_0 h, \quad y_1 + y_1/r_0 h, \quad z_1 + z_1/r_0 h.$$

The position vectors $\mathbf{MT}$, $\mathbf{MP}$, $\mathbf{ME}$ and $\mathbf{MS}$, which we shall denote by $\mathbf{p}$, $\mathbf{q}$, $\mathbf{s}$, $\mathbf{t}$, respectively, have the following projections:

$$\mathbf{p} = (x, y, z), \tag{4.2}$$

$$\mathbf{q} = [x_1(1 + h/r_0), \quad y_1(1 + h/r_0), \quad z_1(1 + h/r_0)], \tag{4.3}$$

$$\mathbf{s} = (0, 0, R), \tag{4.4}$$

$$\mathbf{t} = (x_1, y_1, z_1). \tag{4.5}$$

These vectors are related by the equation

$$\mathbf{s} = \mathbf{p} = \kappa(\mathbf{s} - \mathbf{q}). \tag{4.6}$$

where κ is a scalar quantity. It is also true that

$$\mathbf{q} = \mathbf{p} + \mathbf{TS} + \mathbf{SP}, \tag{4.7}$$

where

$$\mathbf{TS} = \delta\mathbf{p} = (\delta x, \delta y, \delta z),$$
$$\mathbf{SP} = \delta\mathbf{q} = (x_1/r_0 h, y_1/r_0 h, z_1/r_0 h). \tag{4.8}$$

In addition,

$$\delta\mathbf{q} = \frac{(\mathbf{p} + \delta\mathbf{p})}{|\mathbf{p}|} h \tag{4.9}$$

and, hence,

$$\mathbf{s}(1 - \kappa) = \mathbf{p} - \kappa\mathbf{p}(1 + h/|\mathbf{p}|) - \kappa \, \delta\mathbf{p}(1 + h/|\mathbf{p}|). \tag{4.10}$$

Therefore, $\delta\mathbf{p}$ will be known if the scalar κ is determined. This can be done by employing the relation

$$|\mathbf{p}| = |\mathbf{p} + \delta\mathbf{p}| = 1. \tag{4.11}$$

which means that

$$\mathbf{p} \cdot \delta\mathbf{p} \cong 0. \tag{4.12}$$

By multiplying the vector relation (4.10) by $\mathbf{p}$, we eliminate $\delta\mathbf{p}$ from it, and the resulting relation allows us to determine κ. We find that

$$\kappa = \frac{\mathbf{p}(\mathbf{p} - \mathbf{s})}{\mathbf{p}(\mathbf{p} - \mathbf{s}) + h}, \tag{4.13}$$

and relation (13.21) yields for $\delta\mathbf{p}$ the expressions

$$\delta\mathbf{p} = \frac{h}{h + |\mathbf{p}|} \frac{\mathbf{s} - (\mathbf{s} \cdot \mathbf{p}) \cdot \mathbf{p}}{\mathbf{p}(\mathbf{s} - \mathbf{p})}. \tag{4.14}$$

Let us now denote by $\mathfrak{G}$ the matrix transforming the frame $Mxyz$ to the frame $M\xi\eta\zeta$. Its explicit form is

$$\mathfrak{G} = \begin{pmatrix} \cos\theta \cos l - \sin\theta \sin b \sin l & -\sin\theta \cos l - \cos\theta \sin b \sin l & \cos b \sin l \\ \sin\theta \cos b & \cos\theta \cos b & \sin b \\ -\cos\theta \sin l - \sin\theta \sin b \cos l & \sin\theta \sin l - \cos\theta \sin b \cos l & \cos b \cos l \end{pmatrix} \tag{4.15}$$

The components of the vector $\mathbf{p}$ in the standard frame of reference are then

$$\mathbf{p} = (\xi, \eta, \zeta) \tag{4.16}$$

whereas those of vector $\mathbf{t}$ are

$$\mathfrak{G}\mathbf{t} = (\xi_1, \eta_1, \zeta_1), \tag{4.17}$$

i.e.,

$$\mathfrak{G}\delta\mathbf{p}=(\mathbf{t}-\mathbf{p})=(\delta\xi,\,\delta\eta,\,\delta\zeta). \tag{4.18}$$

All three coordinates of the point T can thus be directly measured from the plate, and from them the coordinates of the point S can be deduced after h has been determined. To do this, we need another plate exposed at much different libration angles. We shall use the subscript 1 for quantities referring to this plate. Thus, let l_1, b_1 be the libration angles of this plate, where $|l-l_1|, |b-b_1|$ are as large as possible. Let, also, $\mathbf{p}_1$ and $\mathbf{q}_1$ be the position vectors of the points T_1 and P_1 which correspond to T and P, respectively. If we treat $l-l_1$ and $b-b_1$ as small and call them δl and δb, then we can write that

$$\mathbf{p}-\mathbf{p}_1=\delta\mathbf{p}_1, \quad \mathbf{q}-\mathbf{q}_1=\delta\mathbf{q}_1, \tag{4.19}$$

and

$$\delta\mathbf{p}_1=\left(\frac{\partial\mathfrak{G}}{\partial l}\,\delta l+\frac{\partial\mathfrak{G}}{\partial b}\,\delta b\right)\mathbf{p}, \tag{4.20}$$

$$\delta\mathbf{q}_1=\left(\frac{\partial\mathfrak{G}}{\partial l}\,\delta l+\frac{\partial\mathfrak{G}}{\partial b}\,\delta b\right)(\mathbf{p}+\delta\mathbf{p})\,(1+h/|\mathbf{p}|). \tag{4.21}$$

In the second plate, the conical projection of the point P_1 will not coincide with the conical projection of T_1 as was true (or taken by definition) in the first plate. Let us call T' the intersection of the mean sphere by the line EP_1, and $\mathbf{p}'$ the selenocentric position vector corresponding to it. This vector can be directly obtained from plate measurements, and expressed as $\mathfrak{G}\mathbf{p}'$ in the standard coordinate system. Thus the equation of condition will be of the form

$$\mathfrak{G}(\mathbf{p}'-\mathbf{p})+\left(\frac{\partial\mathfrak{G}}{\delta l}\,\delta l+\frac{\partial\mathfrak{G}}{\partial b}\,\delta b\right)[\mathbf{p}'-\mathbf{p}(1+h/|\mathbf{p}|]$$

$$=(1+h/|\mathbf{p}|)\left(\frac{\partial\mathfrak{G}}{\delta l}\,\delta l+\frac{\partial\mathfrak{G}}{\delta b}\,\delta b\right)\delta\mathbf{p} \tag{4.22}$$

in which h and $\mathbf{dp}$ are not known. However, by post-multiplication of both sides by

$$\mathbf{p}^r\cdot\left(\frac{\partial\mathfrak{G}}{\partial l}\,\delta l+\frac{\partial\mathfrak{G}}{\partial b}\,\delta b\right)^{-1}$$

we eliminate the right-hand side of Equation (4.22) and obtain

$$\mathbf{p}^r\left(\frac{\partial\mathfrak{G}}{\partial l}\,\delta l+\frac{\partial\mathfrak{G}}{\partial b}\,\delta b\right)^{-1}\mathfrak{G}(\mathbf{p}'-\mathbf{p})+\mathbf{p}^r[\mathbf{p}'-\mathbf{p}(1+h/|\mathbf{p}|)]=0, \tag{4.23}$$

where $\mathbf{p}^r$ is the transposed matrix (row vector) or $\mathbf{p}$. Equation (4.23) is scalar and can be solved for h; in doing so we find that

$$h=|\mathbf{p}|^{-1}\left[\mathbf{p}^r\left(\frac{\partial\mathfrak{G}}{\partial l}\,\delta l+\frac{\partial\mathfrak{G}}{\partial b}\,\delta b\right)^{-1}\mathfrak{G}(\mathbf{p}'-\mathbf{p})+\mathbf{p}^r(\mathbf{p}'-\mathbf{p})\right]. \tag{4.24}$$

Once h is computed, a substitution in Equation (4.23) will give for $\delta\mathbf{p}$ the vector equation

$$\delta\mathbf{p}=\frac{1}{(1+h/|\mathbf{p}|)}\left[\left(\frac{\partial\mathfrak{G}}{\partial l}\delta l+\frac{\partial\mathfrak{G}}{\partial b}\delta b\right)^{-1}\mathfrak{G}(p'-p)-[p'-p(1+h/|\mathbf{p}|)]\right].$$

(4.25)

The last two formulae can be generalized by dropping the assumption concerning the size of δl and δb; and, in such a case, they become

$$h=|\mathbf{p}|^{-1}\left[\mathbf{p^{r}}(\mathfrak{G}_{1}-\mathfrak{G})^{-1}\mathfrak{G}(\mathbf{p}'-\mathbf{p})+\mathbf{p^{r}}(\mathbf{p}'-\mathbf{p})\right],$$

(4.26)

and

$$\delta\mathbf{p}=\frac{1}{(1+h/|\mathbf{p}|)}\left[(\mathfrak{G}_{1}-\mathfrak{G})^{-1}\mathfrak{G}(\mathbf{p}'-\mathbf{p})-[\mathbf{p}'-\mathbf{p}(1+h/|\mathbf{p}|)]\right].$$

(4.27)

Thus, two plates of the Moon exposed at widely different libration angles are sufficient to calculate both h and $\delta\mathbf{p}$ and from them the absolute coordinate of the point P. However, the size of the differential displacement of objects situated off the 'mean sphere' is very small and, hence, repeated measurements of many photographs must be used to improve the accuracy of the results.

One essential addition to the above is the correction of the error originating from the variation with the libration in latitude of the center M of the visible disc. This correction must take the form of a translational term in transformation (4.16). If $\mathfrak{H}$ is the three-by-one correcting matrix, then Equation (4.16) should read

$$\mathfrak{G}\mathbf{p}+\mathfrak{H}=(\xi,\eta,\zeta);$$

(4.28)

whereas, for the second plate, the equivalent transformation must be

$$\mathfrak{G}_{1}\mathbf{p}_{1}+\mathfrak{H}_{1}=(\xi_{1},\eta,\zeta_{1}),$$

(4.29)

where $\mathfrak{H}_{1}\neq\mathfrak{H}$.

The basic elements of this stereoscopic method were employed already by the heliometric observers to triangulate the distance of the crater Mösting A (see the preceding chapter) from the Moon's centre. Such observations furnished, in addition to the coordinates λ and β as given by Equations (3.18), also the angular distance h between the Moon's centre and Mösting A at the mean distance of the Moon to be equal to $932''.98\pm0''.19$ m.e. (Koziel, 1967a, b), corresponding to the local radius-vector of 1738.64 km with an uncertainty of ±0.35 km (m.e.). Observations of other points on the Moon made to put the stereoscopic theory to task were initiated by Franz (1901), Saunder (1905, 1907, 1911); and by other investigators whose work we shall proceed to review.

The catalogue of Saunder lists the positions of 1433 features on the lunar surface. These were measured on plates taken with the 24-in. equatoreal coudé of the Paris Observatory, 1895–99. They are gibbous photographs with both morning and evening terminators. For example, the first two plates measured were taken shortly after first quarter and shortly before last quarter. Therefore, the two plates have only a small area of the central portion of the lunar disc in common.

Plate constants for the first two positive plates were derived from a large number of points measured at the telescope. These included Mösting A and the eight fundamental points measured with the heliometer by J. Franz and nineteen features measured by Saunder. The latter were micrometric measurements made with a filar micrometer. Two of the fundamental points of Franz, Aristarchus and Byrgius A, were deleted because of difficulty with their measurement on the plates. Plate constants for the negative plates included the determination of three additional features by Barnard with the 40-in. Yerkes refractor.

Measurements were made using a reseau and an astrographic micrometer with a scale in the eyepiece. The first attempt was to record the réseau on a positive copy of the lunar image. However, too much detail was lost in constructing the positive and as a result the reseau was clamped over the negatives. Each feature was measured four times, twice in opposite plate orientations.

At a later date, two plates taken with the 40-in. refractor at the Yerkes Observatory (1901) were measured by Saunder. These had been taken by G. W. Ritchey in an experiment to determine the detail which can be photographed under the best atmospheric conditions. Unfortunately, there is some question as to the actual dates of these photographs, which is necessary to lunar control studies. Saunder's analysis indicates that the listed date of one photograph is in error by one month and the other by one day. This casts considerable doubt on the value of these observations.

In all, 2885 points were measured and reduced on the Yerkes plates. These included the points previously observed on the Paris plates. This work served as the main basis for the catalogue of IAU lunar coordinates by Blagg and Müller (1935).

The 150 Moon craters measured by Franz are dispersed over the earthward hemisphere. Measurements were made on five plates taken with the 36-in. refractor at the Lick Observatory, 1890–1891. These are primarily near full Moon photographs and features were selected that stand apart from their surroundings due to their brightness. Although Franz referred to them as Moon craters, a few bright mountain peaks are included in the list.

Plate constants were determined by Mösting A and the eight additional fundamental points that Franz derived with the heliometer. Measurements were made with an instrument built by Repsold for the Royal Prussian Academy of Sciences. It could only measure one coordinate precisely along the principal scale and this required that the plate be rotated. The X coordinate was measured twice in the 0° and 180° plate orientation, while the Y coordinate was measured with the plate rotated to 90° and 270°. Because of the difficulty in measuring, Franz divided each plate into nine sectors that could be measured in one setting. Along with the features in each sector, the fundamental points were also measured. Thus, each sector was referenced to the lunar coordinate system as described by these fundamental points.

The reduction of the 150 Moon craters was made by Schrutka-Rechtenstamm (1958). This included a new computation of the eight fundamental points measured with the heliometer. The major basis for this new reduction was a better expression of the physical libration than was available to Franz. Also, Schrutka converted the sector measurements into a single set of positions for each plate. The measurements of Franz, as reduced by Schrutka, have been used as a basis for numerous following lunar control.

The study of Baldwin (1963) produced a catalogue of the coordinates for 696 features. These points were measured on five photographs taken with the 36-in. refractor at the Lick Observatory. Two of these plates are near first and last quarter, two others are crescent phase with morning and evening terminators, and one is at gibbous phase. Also, two of these plates are at almost identical librations and their measurements are correlated with regard to the stereographic reduction method.

This array of phase photography does not allow for multiple observations of individual features and the majority are measured on only two plates, the minimum requirement for a single determination of three-dimensional coordinates. The rest of the features are measured on two combinations of three different plates. Unfortunately, one of these combinations contains the two plates of almost identical libration.

Plate constants were derived using the 150 points of Franz as reduced by Schrutka. Due to the nature of the photography, different groups of points were used for each plate. As there were a relatively small number of points common to both systems, it was necessary to make small auxiliary corrections to the plate constants. The final computation of lunar devations was according to the scheme developed by Saunder.

The U.S. Army Map Service (AMS) Lunar Control System consists of two separate catalogues, AMS 1964 and Group NASA 1965. The AMS 1964 system lists the coordinates of 256 features dispersed over the Moon's earthward hemisphere, while Group NASA lists 496 features concentrated in two zones or belts. These areas are 10° north and south of the equator and 10° east or west of the lunar prime meridian.

The number of control features in both of these catalogues was later reduced during their inclusion in the DOD Selenodetic Control System 1966. Two craters were removed from AMS 1964 because their derived heights exceeded 10 km. A third fundamental adjustment was performed on Group NASA which deleted 805 observations equations and reduced the number of craters to 484.

Approximately the same set of photographs were used for both reductions. These were short exposures taken with the 36-in. refractor at Lick Observatory between 1936 and 1945. A great variety of phase angles are present including crescent and gibbous, with morning and evening terminators. A total of 19 plates were used for both reductions, 15 for AMS 1964 and 18 for Group NASA. Of these, 14 plates were common to both reductions.

Measurements were made by sessions rather than for an entire plate. A session refers to the measurement of a group of features in one day. In this manner, different groups of features appearing on a single plate were measured in various sessions. The AMS 1964 system was measured on 15 plates in 32 sessions, while the Group NASA contained 18 plates measured in 131 sessions. No attempt was made to convert the sessions on a plate into a single relative array of measured coordinates.

Therefore, plate constants as such were not determined. Instead, triangulation, rotation, and scale were derived for each session and a single plate may have several different values for these constants. The method of determining translation, rotation, and scale is by a least-squares adjustment between the measurements of features in a session and their IAU coordinates (Blagg and Müller, 1935).

The ACIC Selenodetic System (1965) lists coordinates for approximately 900 features dispersed over the Moon's earthside hemisphere. Of these, 196 are primary positions and about 700 are supplementary or control extension points. A major purpose of this work was to furnish horizontal and vertical control for the ACIC lunar charting effort. These are the 1:1 000 000 scale LACs and the 1:500 000 scale AIC charts.

The primary control net was measured on near full Moon photography taken at the Pic du Midi Observatory and the Naval Observatory at Flagstaff, Ariz. There were eight differently librated observations, of which seven are from Pic du Midi and one from the Naval Observatory. The Pic du Midi photography consists of a sequence of five short exposures, covering a very short period of time, for each observation. The Naval Observatory photography sets consist of three long exposures taken about one minute apart. All of the 196 primary points were measured on each plate in every sequence.

Plate constants were derived by a least-squares transformation between the plate measurements and the projected positions of three selected features from the work of Schrutka-Rechtenstamm. Some of these coordinates were amended by the measured observations to derive a relatively consistent set.

The control extension was measured primarily on sequences of phase photography from Pic du Midi, along with some long exposures from the Naval Observatory. Generally, three sequences of photographs having approximately the same phase angle were used. These were selected so as to present the largest librational between the three sequences. A select group of points were measured on every plate and features in the primary control net were used to develop plate constants.

Craters from 3 to 20 km in diam were measured, with the majority being less than 10 km in diam. Most of the feature coordinates were determined from two or more sets of differently librated sequences. The control extension points are considerably denser in the equatorial region to support more intensive mapping requirements in this area.

The positions of 734 points of the DOD Selenodetic Control System (1966) are listed in this catalogue, which is a combination of the ACIC Selenodetic System (1965) and AMS Lunar Control System (1964) including Group NASA points. The method of reduction was basically the same as used in developing the AMS control. However, the DOD Selenodetic System does not provide an optimum combination of the ACIC and AMS control works.

The Kiev Lunar Triangulation (1967) study resulted in a 'Catalogue of Seleno-centric Positions of 500 Basic Points on the Moon's Surface'. In this context, seleno-centric refers to the center of mass and not the center of figure. This study attempts to transform the origin of lunar positions to the center of mass after they have been determined in the normal manner. It is predicated on the fact that the librations actually occur about the center of mass and not the center of figure.

The observations were made with two different instruments. One was the astrograph (5.5 m focal length) of the Main Astronomical Observatory at Goloseyevo. This instrument has an automatically moving plate holder which is used to obtain long exposures of 10 to 15 s. The purpose was to average photographically the uncorrelated trembling of the images caused by atmospheric turbulence. The second telescope is a 26-in. refractor of 10.5 m focal length at the Pulkovo Observatory. Often two or three photographs were taken at the same librational position and their measurements are combined.

The first effort was to develop a composite catalogue of the selenocentric coordinates of 160 basic points. These points were measured on 16 near full-moon plates taken at Goloseyevo and Pulkovo. They include the 150 features constants were derived by a comparison of the measurements with the positions of ten features of the Schrutka catalogue. A composite catalogue was derived for the three different sources.

After these coordinates have been determined, corrections are derived to translate the origin from center of figure to the center of mass. Then, corrections are developed to convert the surface positions to 'selenocentric' values. Since this adjustment depends on a feature's location, corrections were determined for local areas. The results of this study were used to establish the selenocentric coordinates of 500 basic points.

In the Manchester Selenodetic Control System (1967), 906 features were measured on near full-moon photography taken with the 24-in. equatoreal coudé at Pic du Midi Observatory (1960–1966). There were 18 differently librated observations used in determining lunar positions. In the same manner as the ACIC control study, a sequence of short exposures were measured for each observation to reduce the effects of seeing displacements. This study used six exposures for each observation.

Features were selected for measurement that were small (5–6 km) and could be identified on full-phase photography. These included splash craters, mountain peaks, and other albedo points. The use of this type of photography was to eliminate the false positioning of a feature caused by different solar altitudes (phase effect). The actual measurements were primarily made with a Zeiss coordinate measuring instrument which has a reversible prism. This allowed a feature to be measured in the forward and reverse orientation without rotating the plate (see Figure 4.2).

Plate constants were developed by a least-squares transformation of the measured coordinates to the projected positions of known points. A higher order transformation is used to derive translation, rotation, scale and reduce the effects of atmospheric refraction. Terms beyond the first order are not used when the Moon is photographed at small zenith distances. In all, 41 positions were used to determine plate constants. These were taken from the catalogues of Schrutka (1958) and ACIC (1965).

The 1972 publication of this system is based on essentially similar observational data; and revised values for 700 points were published in an Appendix to Moutsoulas (1972).

The Tucson Triangulation System of 1968 resulted in a catalogue that lists the position of 1355 features on the Moon's earthward hemisphere. It combines the observations of three other control studies along with measurements of plates taken with

Fig. 4.2.   The late Dr Geoffrey A. Mills (1937–1970), center, assisted by W. V. Garner (right), at work on the Manchester selenodetic control system (1967) with the Zeiss comparator in the measuring room of the department of astronomy, University of Manchester.

the 40-in. refractor at Yerkes Observatory. These studies are the works of Saunder, ACIC, and Gavrilov (Kiev Triangulation).

This study consisted of two distinct operations. The first was the determination of the positions of 48 features as measured on 25 star-trailed Yerkes plates. Star trails and the position of Mösting A. (Koziel, 1967) are used to determine orientation and translation that was independent of the heliometer observations of Franz. However, it was still necessary to derive scale from this system.

After the lunar image has been exposed, the telescope's drive is turned off and the trail of a star is recorded. This is used to determine celestial direction. A much larger group of star-trailed plates were to be used, but some difficulty developed in the processing. Normal lunar photographs were taken on the same nights as star-trailed plates and all were processed in the same manner. Since the star-trailed plates were open to the sky for a longer period of time, they were somewhat sky fogged and required a maximum contrast development. This was not done and rather poor quality images resulted.

The second operation was to determine the positions of the 1355 features from 37 differently librated observations. These included three Yerkes plates, six from Saunder, six from Gavrilov, and 22 from ACIC. It should be noted that the Gavrilov and ACIC systems use multiple plates per observation. Therefore, this work combines the measurements of the equivalent of 131 plates. The 48 positions from the star-trailed plates were used to establish translation, rotation and scale.

The ACIC Positional Reference System (1969) was designed to support basic small scale mapping of the lunar farside. It was not intended to constitute a lasting selenodetic work and a catalog of positions was never produced for the system. Its positional results are best recorded in the Lunar Farside (LMP-2) and Polar Charts (LMP-3), scale 1:5 000 000 and Lunar Planning Charts (LOC), scale 1:2 750 000, which will be described in subsequent chapters of this book.

Essentially, the Positional Reference System extended the ACIC Selenodetic System of 1965 to the lunar farside through a system of overlapping perspective projections keyed to Lunar Orbiter Mission photographs and based on parameters provided by Mission Photo Support Data.

Twenty-four 3-in. focal length Lunar Orbiter Mission photographs were selected which approximately encircle the Moon in polar and equatorial bands. The individual photographic segments (framelets) were precisely reassembled to calibrated values of the Lunar Orbiter film réseau. Perspective projections were computed for each selected photograph, based on camera focal length, position, and orientation. ACIC Selenodetic System (1965) points were plotted on the prepared projections in lunar nearside and limb areas. Each perspective projection was fitted to its reassembled base photograph (see Figure 4.3) considering the fit to Selenodetic System (1965) points and the tie between overlapping photographs. The fitting of projections proceeded from the nearside to limb areas and by extension to the lunar farside where a join was effected in the central farside region with an indicated accuracy of 13 km.

The graphic best fit technique employed in the development of the 1969 system sought the minimization of residual errors from spacecraft ephemeris, earthbased control points, camera orientation values and uncorrected photographic distortions. As might be expected, some differences exist in the positions defined by individual photographs and Positional Reference System values are a mean of these differences. Positions for lunar areas not covered by the polar and equatorial bands were obtained by the fitting of additional photography to these bands.

Some means for an independent evaluation of Positional Reference System results have been afforded by Apollo Landmark Tracking Data. These positional determinations employ Apollo astronaut observations of selected lunar landmark features through an onboard sextant, allowing computation of feature positions through relation to the Apollo spacecraft orbital ephemeris. Ten landmark tracking observations in the equatorial region of the lunar nearside, reflect a standard deviation of 2.9 km relative to earthbased telescopic control data. Six landmark tracking observations in the central lunar farside area show a standard deviation of 9.3 km with respect to Positional Reference System 1969 based values.

What did all this work, based on stereo approach, disclose on the shape of the lunar globe? A harmonic analysis of the surfaces defined by these independent sets of data, and a systematic study of their deviations from the mean sphere was in recent years pioneered by Goudas (1963, 1964a, b, c: 1965a, b, c, 1966 a, b). In doing so, he assumed the actual lunar surface to be expansible in a series of tesseral harmonics of the form

$$r(\alpha, \lambda, \beta) = \alpha \left\{ 1 + \sum_{i=0}^{j} \sum_{j=0}^{\infty} \left[J_{i,j} \cos i\lambda + K_{i,j} \sin i\lambda \right] P_j^i(\beta) \right\}, \qquad (4.30)$$

where $P_j^i(\beta)$ stands for the associated Legendre polynomials in the latitude of order j and index i; and $J_{i,j}$, $K_{i,j}$ are numerical coefficients to be determined from the observations. By known properties of surface harmonics, $P_j^i = 0$ for $i > j$; and in order to facilitate the solution. Goudas assumed that $J_{2j-1}^i = K_j^{2i} = 0$ for $i, j = 1, 2 \ldots$, implying

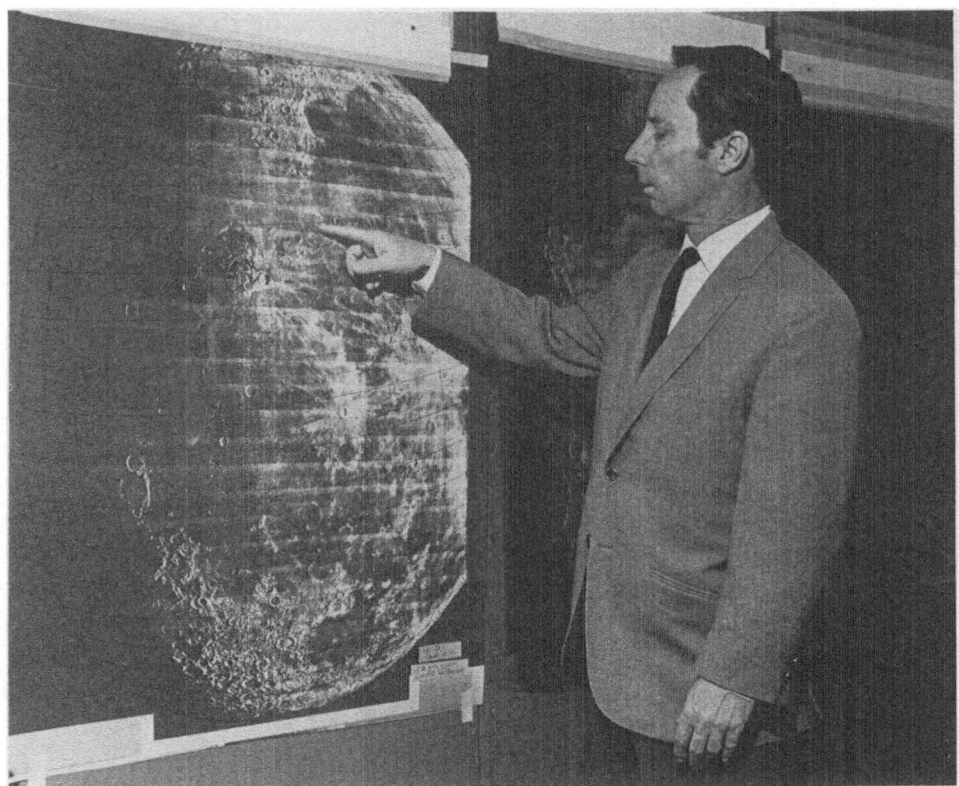

Fig. 4.3.   ACIC cartographer Byron Ruffin demonstrates the fitting of a perspective projection to one of the Lunar Orbiter photographs.

the geometrical similarity of the near and far side of the Moon. With the local values of $r(\alpha, \lambda, \beta)$ given by the stereoscopic method for points of known positions λ and β, Equation (4.30) becomes linear in the coefficients J_j^i, K_j^i, and can be solved for them by the method of least squares. As many equations of condition of the form (4.30) are available for this purpose as are points for which absolute elevations have been established. This is, in general, several hundred (196 for ACIC, 734 for DOD-66, or 917 for the Manchester system), and the corresponding work must be performed with the aid of electronic computers.

TABLE 4.1

The coefficients $10^5 J_{i,j}$ and $10^5 K_{i,j}$ of lunar surface deformation

Coefficient	ACIC	DOD-66	Manchester
$J_{0,2}$	-0.59	0.24	-0.772
$J_{2,2}$	0.36	0.23	0.253
$J_{0,3}$	0.41	-0.64	0.771
$J_{2,3}$	-0.059	0.071	0.0885
$J_{0,4}$	0.11	-0.06	0.487
$J_{2,4}$	-0.060	-0.069	-0.00789
$J_{4,4}$	0.0068	0.0062	0.00337
$J_{0,5}$	0.41	-0.05	0.144
$J_{2,5}$	-0.0031	0.0016	-0.0291
$J_{4,5}$	0.00032	0.00041	-0.000663
$J_{0,6}$	0.38	0.90	0.200
$J_{2,6}$	-0.0060	0.0056	-0.00662
$J_{4,6}$	0.00027	0.00021	-0.000574
$J_{6,6}$	-0.000027	-0.000012	-0.000017
$J_{0,7}$	-0.39	0.66	-0.0125
$J_{2,7}$	0.010	0.0013	0.00272
$J_{4,7}$	-0.000022	0.0000004	0.000303
$J_{6,7}$	-0.000006	-0.0000096	-0.000002
$J_{0,8}$	0.085	0.21	0.0291
$J_{2,8}$	-0.0025	-0.0041	0.00643
$J_{4,8}$	0.00013	-0.0000052	0.000222
$J_{6,8}$	-0.0000012	0.0000032	0.000001
$J_{8,8}$	0.00000014	0.0000002	0.000001
$K_{1,1}$	-0.43	-0.27	-0.199
$K_{2,1}$	0.59	0.31	0.322
$K_{3,1}$ $-$	0.08	0.29	0.000443
$K_{3,3}$	-0.0081	-0.11	0.00942
$K_{4,1}$	0.28	0.24	0.453
$K_{4,3}$	0.017	-0.0068	0.000859
$K_{5,1}$	-0.25	-0.22	-0.147
$K_{5,3}$	0.0031	-0.0012	-0.000084
$K_{5,5}$	0.00029	0.00033	-0.000065
$K_{6,1}$	-0.079	-0.19	-0.0396
$K_{6,3}$	-0.00027	-0.0022	-0.00118
$K_{6,5}$	-0.00014	-0.00010	-0.000173
$K_{7,1}$	0.16	0.081	-0.0646
$K_{7,3}$	0.0007	0.0013	-0.000392
$K_{7,5}$	0.000009	0.0000006	-0.000006
$K_{7,7}$	0.0000003	0.0000021	0.000001
$K_{8,1}$	-0.036	-0.059	0.0321
$K_{8,3}$	0.0011	-0.000039	0.000553
$K_{8,5}$	0.000011	0.0000099	0.000027
$K_{8,7}$	-0.00000056	0.0000001	0.000001

The numerical results of this harmonic analysis are given in Table 4.1, incorporating the principal results obtained by the following three groups: ACIC, DOD-66, and Manchester; the harmonic analysis of which was performed by Bray and Goudas (1966) for ACIC and DOD-66; and by Mills with Sudbury (1968) for the Manchester system. A graphic representation of the isohypses defining the deformations of the actual lunar surface from a sphere of 1738 km mean radius, and based on the three independent sets of the data, are shown on Figures 4.4–4.6; while Figures 4.7 and 4.8 give the profiles of lunar cross-sections in the plane of the equator (Figure 4.7) and of the principal meridian (Figure 4.8) according to the data by ACIC and DOD-66.

In any attempt to assess the significance of these results, we should keep in mind that the parallactic displacements due to differences in elevation, caused by lunar

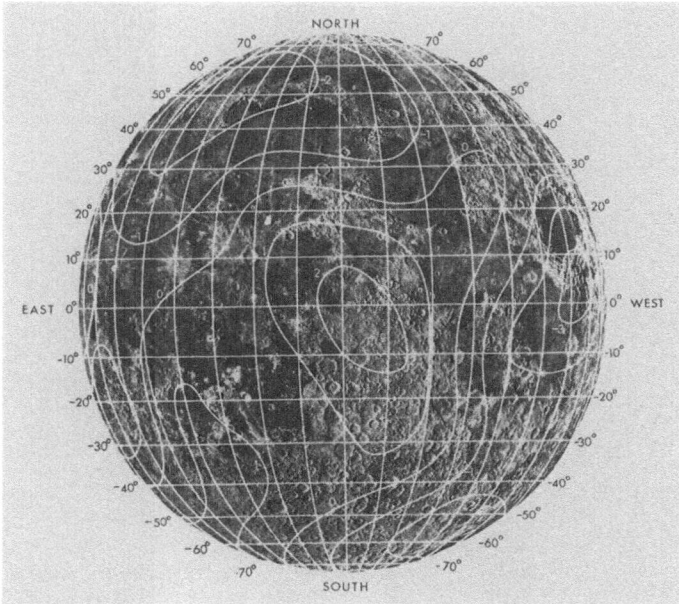

Fig. 4.4.   Isolevel contour map of the Moon, based on the ACIC hypsometric data (after Bray and Goudas, 1966), and projected on the LEM-1B USAF Lunar Mosaic chart. Eight surface harmonics were used in the representation.

librations, are so small that astrometric observations of highest accuracy are required to obtain hypsometric data of any significance; and even the best observations made from the distance of the Earth are subject to errors which render the results of their reductions uncertain within appreciable fractions of their absolute values. Reductions of photographs taken from orbiting spacecraft (Orbiter, Apollo) can, in principle, furnish much more accurate data on the shape of the intersection of their particular orbits (mainly equatorial) with the lunar globe.

Until the advent of the lunar spacecraft, stereoscopic phenomena exhibited in the course of lunar librations afforded the only way to find out anything about the departures of our satellite from spherical shape. The first contributions made by spacecraft to our knowledge of the Moon's shape came from the hard-landing Rangers of 1964–1965. Determinations of the positions in space of points at which the free-flight trajectories of these spacecraft came to their end on crash-landing led to the

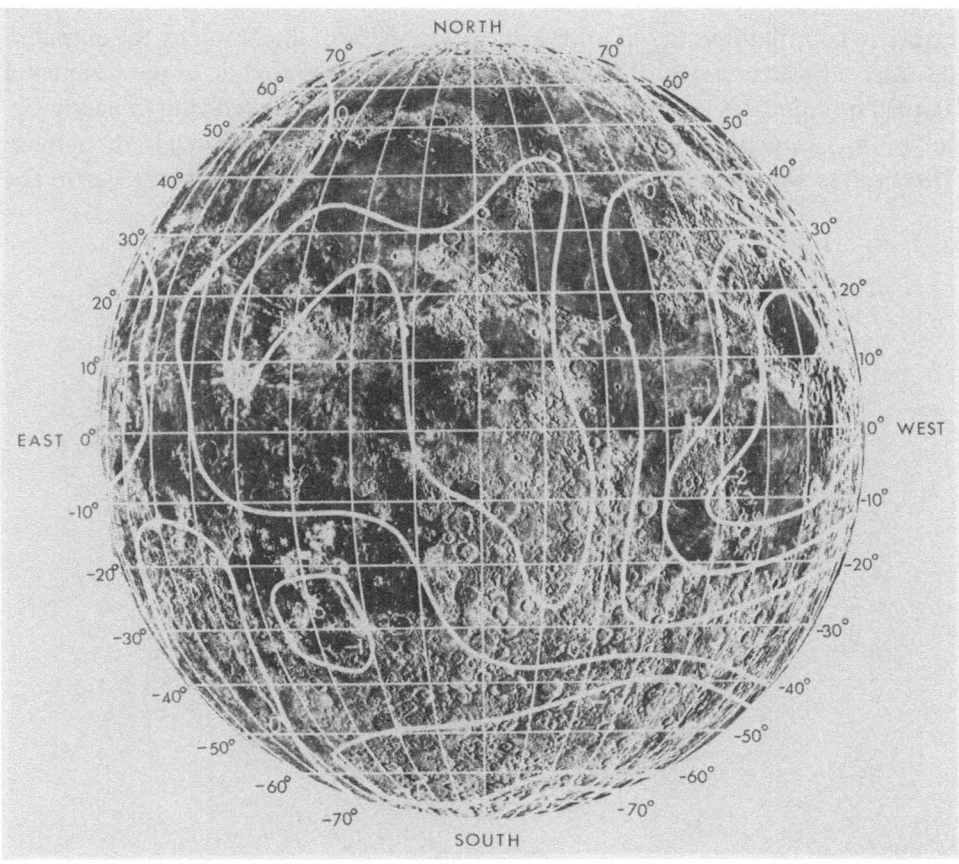

Fig. 4.5. Isolevel contour map of the Moon, based on the DOD-66 hypsometric data (after Bray and Goudas, 1966), and projected on the LEM-1B USAF Lunar Mosaic chart. Eight surface harmonics were used in the representation.

data summarized in Table 4.2 (after Sjøgren, 1967); and furnished four absolute radii of the Moon, connecting the landing points with the centre of mass of our satellite, accurate within ± 100 m.

Hypsometric results of much greater importance were obtained since the advent of the Apollo project between 1969–1972. As is well known, Apollo 11, 12 and 14 missions, in 1969, followed by the Luna 17 and 21, deposited on the lunar surface cube-corner reflectors, capable of returning laser signals flashed from the Earth; and from the timing of the return of such light 'echoes' from the Moon the instantaneous distance between the transmitter on Earth and the reflector on the Moon can be determined within one part in 10^8 (i.e., ± 4 m). Since this distance varies with the time as a result of the continuous relative motion of both stations, long series of observations are needed to specify the absolute elevation of each respective cube-corner. The necessary measurements have been in progress for some time; and although few specific results have been reported so far, they can be confidently expected in the future.

However, Apollo 15 and 16 missions of 1971–1972 have already made much more extensive contributions to our knowledge of the shape of the Moon by the output of the laser altimeters mounted in the scientific instruments module of the Command and Service modules of the respective missions, which were revolving in nearly circular orbits around the Moon while the Excursion Module descended to the surface. These orbits were monitored continuously by the tracking stations on the Earth. The

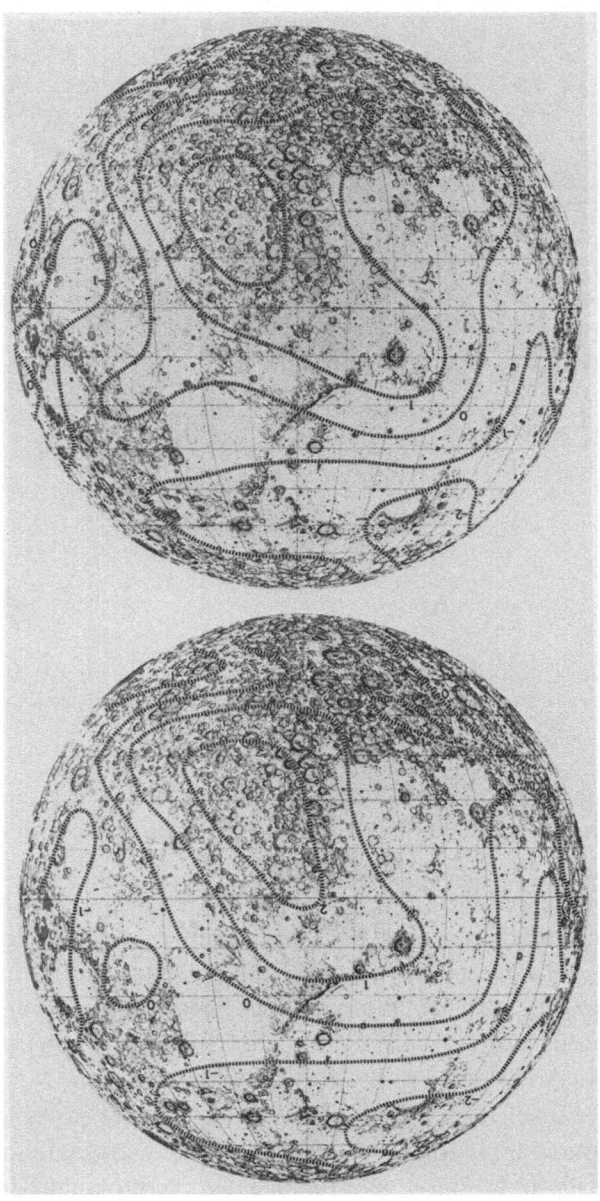

Fig. 4.6.   Isolevel contour map of the Moon, based on the Manchester hypsometric data (after Mills, 1968).

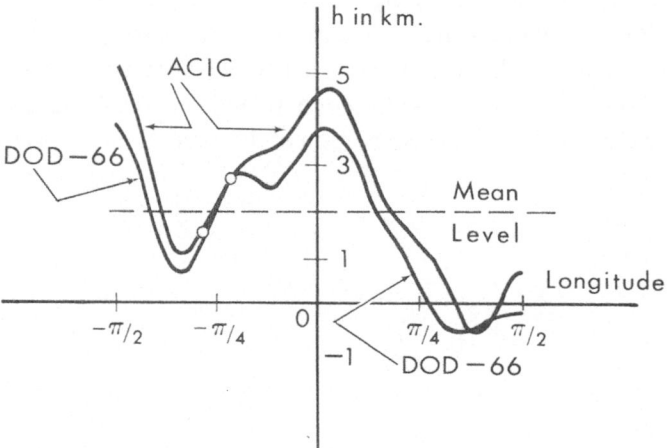

Fig. 4.7.   Equatorial section of the Moon as derived from the systems ACIC and DOD-66.

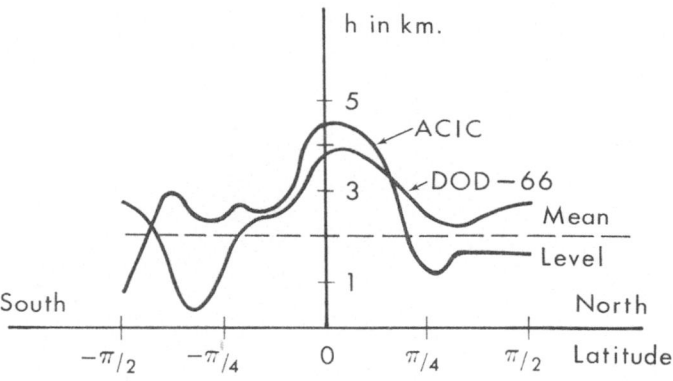

Fig. 4.8.   First meridian section of the Moon as derived from the systems ACIC and DOD-66.

TABLE 4.2

Absolute lunar radii inferred from the impacts of Ranger spacecraft

Spacecraft	Date and time of impact		Coordinates of impact point		Local radius (in km)
Ranger 6	1964 Feb. 2,	$9^h24^m33^s$UT	21°52 W	9°33 N	1735.3
Ranger 7	1964 July 31,	13 25 49	20.58 E	10.63 S	1735.5
Ranger 8	1965 Feb. 20,	9 57 37	24.65 W	2.67 N	1735.2
Ranger 9	1965 Mar. 24,	14 8 20	2.37 E	12.83 S	1735.7

position of the spacecraft relative to the Moon's centre thus being known within a few metres, and its altitude above the actual surface being determined from the time-delay of returning laser echoes with the same accuracy, the difference between the two will furnish the distance of the reflecting element of the surface from the Moon's centre within total errors not exceeding ± 50 to 75 m (i.e., 10 to 100 times smaller than any data previously obtainable from the distance of the Earth).

The altitude profiles of the Moon measured by the laser altimeters of the Apollo 15 and 16 missions are shown on the accompanying Figures 4.9 and 4.10. Each set of these data refers to a different cross-section of the orbit of the respective spacecraft

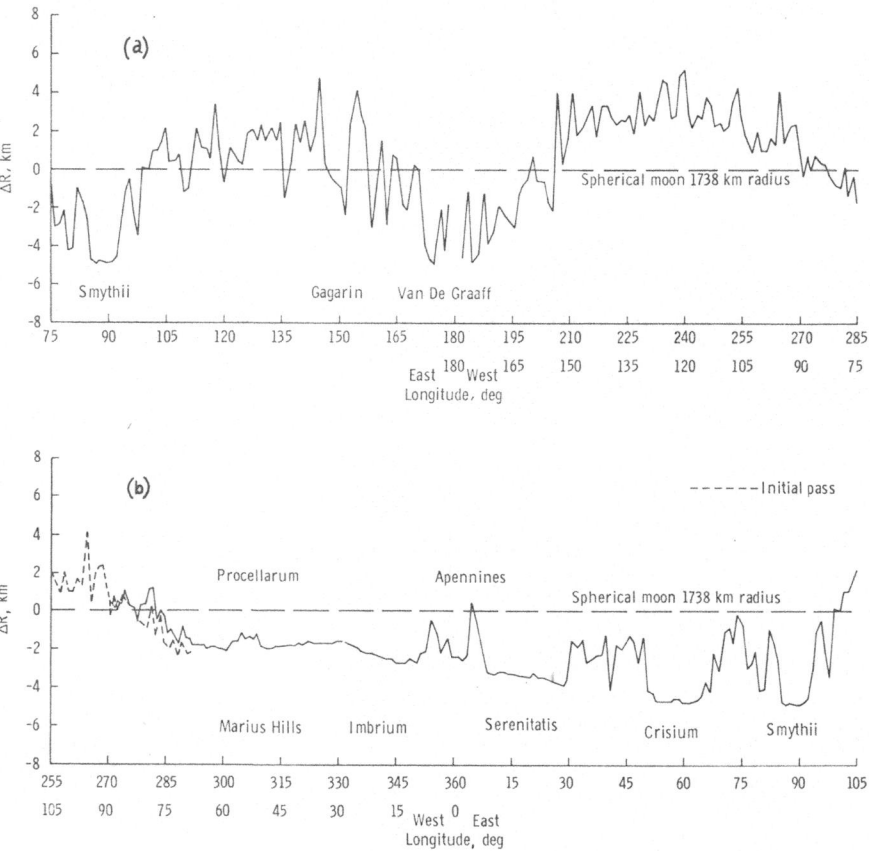

Fig. 4.9.   Apollo 15 altitude profile and radius deviations from spherical Moon. (a) – lunar farside. (b) – lunar nearside.

with the surface of the Moon; and cannot say anything on absolute elevations anywhere else. But they disclose that – along each cross-section – the actual lunar surface deviates from the mean sphere of 1738 km radius in a very complicated manner, not describable by any single (or even a small group of) surface harmonics. The outlines of individual maria or mountain chains stand out clearly on the continental background; and other local features known from surface photography can be distinguished in the records.

Perhaps the outstanding result furnished by the laser altimeters of Apollo 15 and 16 has been the realization that the front hemisphere of the Moon – far from being

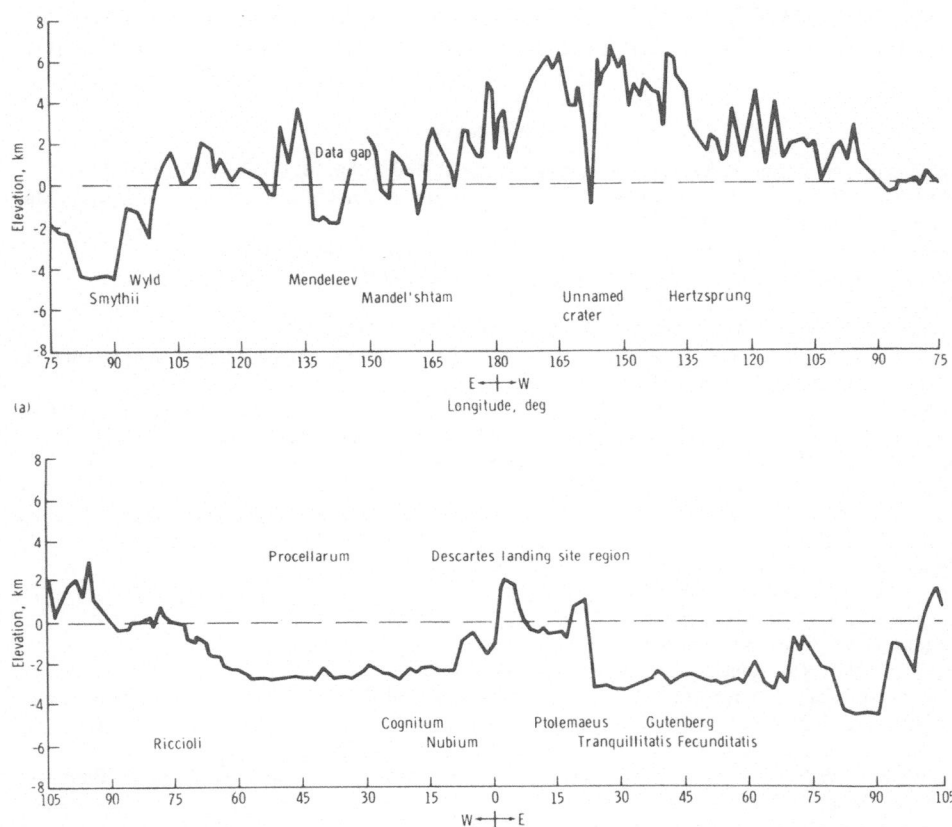

Fig. 4.10.   Apollo 16 altitude profile and radius deviations from spherical Moon of mean radius 1738.0 km.
(a) Lunar far side, (b) near side.

alongated in the direction of the Earth in any kind of 'fossil tide' – is actually depressed below the mean Moon-level; and it is the far side which is elongated. One could conclude from this that the centre of symmetry of the actual figure of the Moon is displaced by 2–3 km from the Moon's centre of mass. Such an interpretation of the observed facts – while possible – is, however, not yet necessary.

Any asymmetry between the front and far side of the Moon, as seen from the Earth, is describable in terms of the *odd* harmonics in selenographic longitude, in which any particular cross-section can be decomposed. A presence, in this harmonic spectrum, of the first harmonic $(j=1)$ would necessitate indeed the centre of figure to be displaced from the centre of mass, to an extent consistent with its amplitude. If, however, the observed asymmetry of figure can be represented by a combination of harmonics higher than the first (corresponding to $j=3$, 5, etc.) with appropriate coefficients, no shift between centre of figure and of mass is required any more to account for the observed facts. The two altitude profiles furnished by Apollo 15 and 16 are insufficient to decide without ambiguity which of the two alternatives may represent better the observed facts; and as no more Apollo flights are scheduled to contribute further to the available evidence, the issue is likely to remain undecided for some time to come.

References

ACIC Selenodetic Control System: 1966; cf. Meyer, D. L. and Ruffin, B. W.: 1965, *Icarus* **4**, 513.

AMS Selenodetic Control System: 1964; cf. Breece, S., Hardy, M., and Marchant, M. Q., U.S. Army Map Service, Tech. Rept. No. 29.

Baldwin, R. B.: 1963, *The Measure of the Moon*, Univ. of Chicago Press, Chapter 11.

Blagg, M. A. and Müller, K.: 1935, *Named Lunar Formations*, Vols. 1 and 2, London.

Bray, T. A. and Goudas, C. L.: 1966, *Icarus* **5**, 526.

Carson, D., Davidson, M. E., Goudas, C. L., Kopal, Z., and Stoddard, L. G.: 1966, *Icarus* **5**, 334.

Davidson, M. E., Goudas, C. L., and Kopal, Z.: 1967, in *Measure of the Moon* (ed. by Z. Kopal and C. L. Goudas), D. Reidel Publ. Co., Dordrecht, pp. 144–175.

DOD Selenodetic Control System: 1966, cf. Eigen, J. M. and Hathaway, J. D., in *Measure of the Moon*, (ed. by Z. Kopal and C. L. Goudas), D. Reidel Publ. Co., Dordrecht, pp. 305–316.

Dommanget, J.: 1962, *Comm. Obs. Roy. Belgique*. No. 208.

Franz, J.: 1901, *Mitt. Sternwarte Breslau* **1**, 1–48.

Gorynia, A. A. and Drofa. V. K.: 1962, *Relief of the Limb Regions of the Moon*, Publ. House Ukrainian Acad. Sci., Kiev.

Goudas, C. L.: 1963, *Icarus* **2**, 423.

Goudas, C. L.: 1964a, b, c. *Icarus* **3**, 168, 273, 275.

Goudas, C. L.: 1965a, b, c. *Icarus* **4**, 188, 218, 528.

Goudas, C. L.: 1966a, b, *Icarus* **5**, 99, 316.

Hayn, F.: 1907, *Abh. König. Sachs. Gesell. Wiss.* **30**, 1–105.

Kiev Lunar Triangulation System: 1967, cf. Gavrilov, I. V. and Kisliuk, V. S., 'Svodnyj Katalog selenocentricheskich polozhenej 2580 basisnych toček na Lunĕ', Naukova Dumka Press, Kiev. 1970.

Koziel, K.: 1967a, *Icarus* **7**, 1–28.

Koziel, K.: 1967b, in *Measure of the Moon* (ed. by Z. Kopal and C. L. Goudas), D. Reidel Publ. Co., Dordrecht, pp. 3–11.

Manchester Selenodetic Control System: 1967, cf. Mills, G. A., *Icarus* **6**, 131; **7**, 193; **8**, 90.

Mills, G. and Sudbury. P. V.: 1968, *Icarus* **9**, 538.

Nefediev, A. A.: 1957, *Bull. Engelhardt Obs., Kazan*, No. 30.

Potter, H. I.: 1962, in Z. Kopal and Z. K. Mikhailov (eds.), 'The Moon', *IAU Symp.* **14**, Academic Press, New York and London, pp. 63–66.

Saunder, S. A.: 1905, *Monthly Notices Roy. Astron. Soc.* **65**, 458.

Saunder, S. A.: 1907, *Mem. Roy. Astron. Soc.* **57**, 1.

Saunder, S. A.: 1911, *Mem. Roy. Astron. Soc.* **60**, 1.

Schimerman, L. A.: 1973, in *NASA Lunar Cartographic Dossier* (ed. by L. A. Schimerman), Vol. 1, Section 3.1.

Schrutka-Rechtenstamm, G. von: 1958, *Sitzungsber. Österr. Akad. Wiss. (Math.-Naturwiss. Klasse)* **167**, 71–106.

Sjøgren, W. L.: 1967, in *Measure of the Moon* (ed. by Z. Kopal and C. L. Goudas), D. Reidel Publ. Co., Dordrecht, pp. 341–343.

Tucson Lunar Triangulation System: 1968, cf. Arthur, D. W. G., 'A New Secondary Selenodetic Triangulation', *Comm. Lunar Planetary Labor.* **7**, 130.

Watts, C. B.: 1963, 'The Marginal Zone of the Moon', *Astron. Papers Amer. Ephemeris and Nautical Almanac*, Vol. 17.

Weimer, Th.: 1952, *Atlas des Profils Lunaires*, Publ. de l'Observ. de Paris.

Wollenhaupt, W. R. and Sjøgren, W. L.: 1972a, *Moon* **4**, 337.

Wollenhaupt, W. R. and Sjøgren. W. L.: 1972b, in *Apollo 16 Preliminary Science Report*, (NASA SP-315), pp. 30–1 to 5.

Yakovkin, A. A.: 1952, *Trans. IAU* **8**, 231.

CHAPTER 5

RELATIVE ELEVATIONS ON THE MOON

In the preceding chapter we outlined the principal sources of our present knowledge of the shape of the Moon, and of the absolute elevations of its individual surface features. To determine such absolute altitudes constitutes a severe and exacting task calling for astrometric work of the highest accuracy (or, in the case of laser ranging, for similarly accurate measurements of light transit times), from which the global shape of the Moon is emerging at the present time. On the other hand, the determination of the *differences* in altitudes of individual points on the Moon *relative* to the adjacent surface constitutes much less difficult or time-consuming task; for several reasons.

First, because any difference in relative altitudes on the Moon causes the more elevated eminence to cast a *shadow* on the surrounding landscape; and this shadow – at low altitude of the illuminating sunlight – magnifies any altitude difference by as much as a factor of 100 near sunrise or sunset. Secondly, to measure the length of the shadow cast in the rays of the rising or setting Sun may call again for angular measurements of highest accuracy, but – and this is significant – no longer of large angles (comparable with the apparent semi-diameter of the Moon), but only of a tiny fraction of it; and this makes all the difference from the instrumental point of view.

In order to ascertain such relative elevations of specific lunar features from the measured lengths of their shadows, let us depart from the geometry as shown on Figure 5.1, with the aim of determining the height h of a sunlit lunar eminence P, casting a shadow whose tip, as seen from the Earth, is situated at another point S. Let the slant range between P and S be denoted by δ, and the distance between S and the observing site O on Earth be r'_s. Let, moreover, the angle θ at S between the vectors $PS \equiv \mathbf{s}$ and $OS \equiv \mathbf{r}'_s$ represent the difference in elevation of the peak P and the

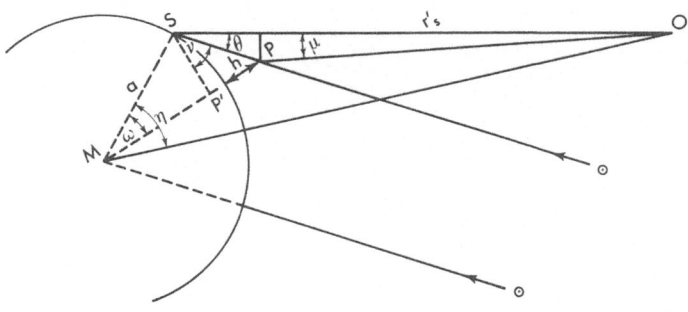

Fig. 5.1.   Geometry of the lunar shadows.

observer O above the lunar horizon. If so, then by solving for $\mathbf{s}$ from the triangle OPS we find that

$$\mathbf{s} = \frac{\mathbf{r}'_s \sin \mu}{\sin(\theta + \mu)}, \tag{5.1}$$

where μ denotes the (topocentric) angular length of the shadow cast by P as seen from O.

Furthermore, if ω denotes the selenocentric length of this shadow (i.e., the angle PMS: cf. again Figure 5.1) and a stands for the mean radius of the Moon, it follows from the triangles involved that

$$a \sin \omega = \mathbf{s} \cos v \tag{5.2}$$

and

$$(a + h) \cos v = a \cos(v - \omega), \tag{5.3}$$

where v is the angle PSP'. Our aim is to solve this latter equation for h. If we divide both sides of it by $a \cos v$, we find that

$$\frac{h}{a} = \frac{\cos(\omega - v)}{\cos v} - 1 = \sin \omega \tan v - 2 \sin^2 \tfrac{1}{2} \omega, \tag{5.4}$$

where from an elimination of $\mathbf{s}$ between (5.1) and (5.2) it readily transpires that

$$\sin \omega = \frac{\mathbf{r}'_s \sin \mu \cos v}{a \sin(\theta + \mu)}. \tag{5.5}$$

Of the auxiliary quantities involved on the right-hand side of the foregoing equation, the topocentric angle μ can be directly measured. The topocentric distance $\mathbf{r}'$ of point S (cf. Figure 5.2) follows from the triangles EMO and MOS (not necessarily co-planar) as a solution of the equations

$$r'^2_s = r'^2 + a^2 - 2ar' \cos \eta, \tag{5.6}$$

$$r'^2 = r^2 + \varrho^2 - 2\varrho r \cos \gamma, \tag{5.7}$$

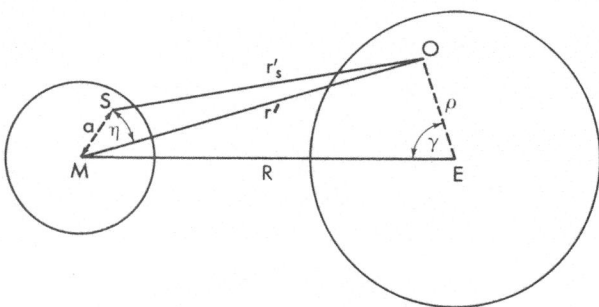

Fig. 5.2.   Geometry of the Earth-Moon configuration.

where $a \equiv MS$ denotes the mean radius of the Moon; $\varrho \equiv OE$, the distance between the center of the Earth and the observing place; and γ, the geocentric angle between the vectors **EO** and **EM**; while η is the selenocentric angle between MO and MS.*
An elimination of r' between (5.6) and (5.7) leads to

$$r_s'^2 = r^2 + \varrho^2 + a^2 - 2a\varrho \cos\gamma - 2a(r^2 + \varrho^2 - 2\varrho r \cos\gamma)^{1/2} \cos\eta =$$
$$= r^2 \{1 + p^2 + q^2 - 2p \cos\gamma - 2q(1 - p \cos\gamma + \tfrac{1}{2}p^2 + \cdots) \cos\eta\}, \qquad (5.8)$$

where we have abbreviated

$$p = \varrho/r \quad \text{and} \quad q = a/r. \qquad (5.9)$$

For the mean values of $r = 384\,400$ km and $\varrho = 6371$ km while $a = 1738$ km, $p = 0.01\,657$ and $q = 0.004\,521$. If we regard these latter ratios small enough for their squares and cross-products to be negligible, the square-root of Equation (5.7) can be approximated by

$$\mathbf{r}_s' = r(1 - p \cos\gamma - q \cos\eta + \cdots) \qquad (5.10)$$

within a small fraction of 1% – with r varying between $364\,400$ and $406\,730$ km in the course of each month on account of the eccentricity of the lunar orbit.

In order to complete the evaluation of the angle ω as defined by (5.5) from the measured values of μ, we must specify γ, η as well as θ and v in terms of observable quantities; and this can be accomplished in the following manner. The geocentric angle γ is (from the triangle EMO; cf. Figure 5.2) defined clearly by

$$\cos\gamma = \sin\delta_{\langle} \sin\varphi + \cos\delta_{\langle} \cos\varphi \cos(\alpha_{\langle} - \Theta), \qquad (5.11)$$

where $\alpha_{\langle}$, $\delta_{\langle}$ denote the Moon's right ascension and declination; Θ, the sidereal time of observation (the difference $\alpha_{\langle} - \Theta = H$ being equal to the lunar geocentric hour angle); and $\varphi_{\langle}$ the geocentric latitude of O – in terms of which the variation of the terrestrial radius-vector ϱ with latitude can be expressed as

$$\varrho = 6378(1 - 0.002\,35 \sin^2\varphi + \cdots) \text{ km} =$$
$$= r \sin\pi_{\langle}(1 - 0.002\,35 \sin^2\varphi + \cdots) \text{ km}, \qquad (5.12)$$

where $\pi_{\langle}$ stands for the equatorial horizontal parallax of the Moon. Similarly, from

$$\cos\eta = \sin b' \sin\beta_s + \cos b' \cos\beta_s \cos(l' - \lambda_s), \qquad (5.13)$$

where b', l' are the topocentric librations of the Moon as defined in Chapter 2 (representing, in effect, the selenographic latitude β_0 and longitude λ_0 of the observer's position at O); and β_S, λ_S, the selenographic latitude and longitude of the point S at which the tip of the shadow is cast on the lunar surface. Moreover, if $A_\odot$, $D_\odot$ denote selenographic coordinates of the Sun as seen from the Moon**, the angle θ between

---

* The reader may note that, in terms of our preceding notations, $\kappa = \eta - \omega$.
** The difference between selenocentric and topocentric values of these coordinates on the Moon becomes immaterial because of the Moon's small size and great distance of the Sun.

the direction to the Sun and the observer at O will be given by

$$\cos \theta = \sin b' \, \sin D_\odot + \cos b' \, \cos D_\odot \, \cos(l' - A_\odot), \; . \tag{5.14}$$

while the angle v equal to the altitude of the Sun above the lunar horizon at P (cf. again Figure 5.1) follows from the equation

$$\sin v = \sin \beta_P \, \sin D_\odot + \cos \beta_P \, \cos D_\odot \, \cos(\lambda_P - A_\odot) \tag{5.15}$$

where β_P, λ_P denote the selenographic coordinates of the point P on the lunar surface.*

With the aid of the auxiliary quantities and angles defined by the foregoing equation, the selenocentric angle ω as given by Equation (5.5) can readily be determined in terms of the measured shadow length μ; and with its aid, the fractional height h/a of the shadow-casting eminence P evaluated from (5.4). An alternative way of determining ω would be to do so from the measured pairs of coordinates β_P, λ_P and β_S, λ_S in place of one of them being combined with μ. If the values of both $\beta_{P,S}$ and $\lambda_{P,S}$ have been determined with the requisite precision, by the method outlined earlier in this section, it follows immediately that

$$\cos \omega = \xi_P \xi_S + \eta_P \eta_S + \zeta_P \zeta_S, \tag{5.16}$$

where $\xi_{P,S}$ etc. are direction cosines of the lunar radii-vectors to the surface points P, S, of the form (3.12); and, consequently,

$$\cos \omega = \sin \beta_P \, \sin \beta_S + \cos \beta_P \, \cos \beta_S \, \cos(\lambda_P - \lambda_S). \tag{5.17}$$

Once the angle ω has thus been determined from known values of β_P, λ_P and β_S, λ_S, and the angle v from (5.15), Equation (5.4) then furnishes at once the desired ratio h/a. This procedure for determining h may, on the face of it, seem more straightforward than the one outlined earlier, which leads to h via the measured (topocentric) shadow length μ. In actual fact this is, however, scarcely the case; for its complexity is stored in the absolute determination of two pairs of coordinates from the measured pair of plane coordinates $x_{P,S}$ and $y_{P,S}$, by a method outlined at the outset of this chapter; and this implies a knowledge of lunar topocentric librations b', l' as well as of the observer's position on Earth.

Throughout all foregoing developments we have tacitly assumed the observer O to be situated on the terrestrial surface and formulated the expression for r'_s in (5.5) accordingly. If, however, the actual observations are made from another vantage point – such as a spacecraft Δ launched to the Moon by human hand – the entire procedure for shadow reductions as outlined in this chapter continues to be valid as it stands, provided that:

(1) the quantity r'_s as given by (5.6) or (5.8) for an Earth-bound observer is replaced

---

* Strictly speaking, these coordinates refer to the point at which the rays from the center of the apparent disc of the Sun are tangent to the shadow-casting obstacle. Should – as may frequently be the case – the profile of this obstacle be convex, the selenographic position of the tangent point may shift somewhat during sunrise or sunset with the varying altitude of the Sun above the lunar horizon. The apparent time displacements of P due to this case are, however, likely in most cases to be too small to be significant.

by the actual distance of the rocket from the point S on the Moon at the time of observation; and that

(2) the selenographic coordinates β_0, λ_0 of the observer at O are no longer identified with the topocentric libration constants b' and l', but replaced in Equations (5.13) and (5.14) by the actual selenographic coordinates β_A and λ_A of the spacecraft at the time of observation.

In order to do so, suppose that the geocentric equatorial coordinates α_A and δ_A of the spacecraft have been obtained from their topocentric values measured at any particular time, and that r_A denotes its instantaneous distance from the Earth's center (obtained from the measured topocentric slant range r_A'). If so, the rectangular co-ordinates of the spacecraft in the geocentric equatorial system will be given by

$$
\begin{aligned}
x_{\oplus A} &= r_{\oplus A} \cos\delta_A \cos\alpha_A, \\
y_{\oplus A} &= r_{\oplus A} \cos\delta_A \sin\alpha_A, \\
z_{\oplus A} &= r_{\oplus A} \sin\delta_A;
\end{aligned}
\tag{5.18}
$$

while the geocentric equatorial coordinates of the Moon's center in the same system are

$$
\begin{aligned}
x_{\oplus\mathbb{C}} &= \mathbf{r} \cos\varrho_{\mathbb{C}} \cos\alpha_{\mathbb{C}}, \\
y_{\oplus\mathbb{C}} &= \mathbf{r} \cos\delta_{\mathbb{C}} \sin\alpha_{\mathbb{C}}, \\
z_{\oplus\mathbb{C}} &= \mathbf{r} \sin\delta_{\mathbb{C}},
\end{aligned}
\tag{5.19}
$$

where $\mathbf{r}$ denotes the radius-vector of the lunar relative orbit.

If so, the *selenocentric* rectangular coordinates, as defined at the outset of this section, will be given by

$$
\begin{aligned}
x_{\mathbb{C}A} = (x_{\oplus A} - x_{\oplus\mathbb{C}}) &(\cos\omega \cos\Omega' - \sin\omega \sin\Omega' \cos i) + \\
+ (y_{\oplus A} - y_{\oplus\mathbb{C}}) &(\cos\omega \sin\Omega' + \sin\omega \cos\Omega' \cos i) + \\
+ (z_{\oplus A} - z_{\oplus\mathbb{C}}) &(\sin\omega \sin i) = r_{\mathbb{C}A} \cos\beta_A \cos\lambda_A,
\end{aligned}
\tag{5.20}
$$

$$
\begin{aligned}
y_{\mathbb{C}A} = (x_{\oplus A} - x_{\oplus\mathbb{C}}) &(-\sin\omega \cos\Omega' - \cos\omega \sin\Omega' \cos i) + \\
+ (y_{\oplus A} - y_{\oplus\mathbb{C}}) &(-\sin\omega \sin\Omega' + \cos\omega \cos\Omega' \cos i) + \\
+ (z_{\oplus A} - z_{\oplus\mathbb{C}}) &(\cos\omega \sin i) = r_{\mathbb{C}A} \cos\beta_A \sin\lambda_A,
\end{aligned}
\tag{5.21}
$$

and

$$
\begin{aligned}
z_{\mathbb{C}A} = (x_{\oplus A} - x_{\oplus\mathbb{C}}) &(\sin\Omega' \sin i) + \\
+ (y_{\oplus A} - y_{\oplus\mathbb{C}}) &(-\cos\Omega' \sin i) + \\
+ (z_{\oplus A} - z_{\oplus\mathbb{C}}) &(\cos i) = r_{\mathbb{C}A} \sin\beta_A,
\end{aligned}
\tag{5.22}
$$

where

$$
r_{\mathbb{C}A}^2 = x_{\mathbb{C}A}^2 + y_{\mathbb{C}A}^2 + z_{\mathbb{C}A}^2
\tag{5.23}
$$

denotes the distance of the spacecraft from the center of the Moon; and ω, the angle between the lunar ascending node and prime meridian (measured along the Moon's equator) can be evaluated from the equations

$$\cos b \ \sin \omega = \{\cos i \ \cos \delta \ \sin(\Omega' - \alpha_{\text{\(}}) - \sin i \ \sin \delta_{\text{\(}}\} \cos l +$$
$$+ \{\cos \delta \ \cos(\Omega' - \alpha_{\text{\(}})\} \sin l, \tag{5.24}$$

or

$$\cos b \ \cos \omega = \{\cos i \ \cos \delta_{\text{\(}} \ \sin(\Omega' - \alpha_{\text{\(}}) - \sin i \ \sin \delta_{\text{\(}}\} \sin l -$$
$$- \{\cos \delta \ \cos(\Omega' - \alpha_{\text{\(}})\} \cos l, \tag{5.25}$$

where the lunar libration angles b, l as well Ω' and i possess exactly the same meaning as in Chapter 2.

Let, furthermore,

$$x_{\text{\(s}} = a \ \cos \beta_s \ \cos \lambda_s,$$
$$y_{\text{\(s}} = a \ \cos \beta_s \ \sin \lambda_s, \tag{5.26}$$
$$z_{\text{\(s}} = a \ \sin \beta_s,$$

denote the selenocentric rectangular coordinates of the shadow tip S on the lunar sphere of mean radius a in the same system. If so, the direction cosines l, m, n of the vector $\mathbf{OS} \equiv \mathbf{\Delta S}$ joining the position of the spacecraft Δ with the point S will be given by

$$l = \frac{x_{\text{\(\Delta}} - x_{\text{\(s}}}{r'_{\Delta s}},$$

$$m = \frac{y_{\text{\(\Delta}} - y_{\text{\(s}}}{r'_{\Delta s}}, \tag{5.27}$$

$$n = \frac{z_{\text{\(\Delta}} - z_{\text{\(s}}}{r'_{\Delta s}},$$

where

$$r'^2_{\Delta s} = (x_{\text{\(\Delta}} - x_{\text{\(s}})^2 + (y_{\text{\(\Delta}} - y_{\text{\(s}})^2 + (z_{\text{\(\Delta}} - z_{\text{\(s}})^2. \tag{5.28}$$

This equation should replace (5.7) for observations made aboard a spacecraft.

Since, moreover, the direction cosines of incident sunlight in the same system of selenographic coordinates are given by

$$l_{\odot} = \cos D_{\odot} \ \cos A_{\odot},$$
$$m_{\odot} = \cos D_{\odot} \ \sin A_{\odot}, \tag{5.29}$$
$$n_{\odot} = \sin D_{\odot},$$

it follows that the angle θ between the radii-vectors from S to Δ and the Sun should be given by the equation

$$\cos \theta = l l_{\odot} + m m_{\odot} + n n_{\odot}, \tag{5.30}$$

which represents a generalization of (5.14) for observations made from the vantage point at Δ – no matter how close to the lunar surface this may happen to be. The selenographic coordinates of a spacecraft at Δ then are given by

$$\sin\beta_\Delta = z_{\langle\Delta} / r_{\langle\Delta} \quad \text{and} \quad \tan\lambda_\Delta = y_{\langle\Delta} / x_{\langle\Delta},\tag{5.31}$$

and its height above the lunar surface,

$$h = r_{\langle\Delta} - a.\tag{5.32}$$

Lastly, the selenographic components of the *velocity* of our spacecraft should follow from a time-differentiation of Equations (5.20)–(5.22) for the selenographic coordinates $x_{\langle\Delta}$, $y_{\langle\Delta}$, and $z_{\langle\Delta}$; care being merely taken to add $+\omega_{\langle}y_{\langle\Delta}$ to $x_{\langle\Delta}$ and $-\omega_{\langle}x_{\langle\Delta}$ to $y_{\langle\Delta}$ as terms arising from the rotation of the Moon with an angular velocity $\omega_{\langle}$ about the z-axis (the selenographic axes rotate).

Whichever method of those outlined above we employ for a determination of the fractional difference h in height between the points P and S by the shadow method, the underlying geometry implies that the accuracy of the results will be maximum at the center of the apparent lunar disc (where the measured shadow length μ or the coordinate differences $x_P - x_S$ etc. are subject to no foreshortening), and diminish progressively toward the limb. Near the limb, both angles θ and μ tend separately to 0; and, as a result, their ratio on the right-hand side of Equation (5.5) for $\sin\omega$ becomes well-night indeterminate – which effectively invalidates the shadow approach to a determination of the heights of lunar mountains in the limb regions. In order to do so, another method must be sought; and this is fortunately made possible by the fact that – on account of libration – the lunar regions within peripheral regions of approximately $\pm 7°$ in width are, from time to time, seen in projection against the dark background of the sky, and their profiles can be accurately measured. Following some early exploratory work by Hayn (1914a, b), a comprehensive atlas of lunar profiles visible at different librations has been published by Weimer (1952) and, more recently, by Nefediev (1958) and Watts (1963). The significance of the outlines of the lunar limb as seen against the bright background of the Sun during annular eclipses has already been pointed out in Chapter 4; while similar silhouettes exhibited by the Moon during partial solar eclipses (cf., e.g., Whitwell, 1929; Fujinami, 1952; Fujinami *et al.*, 1954; Kristenson, 1954, and others) are much more limited in scope and cannot compete with the limb measurements at night.

So much for the situation encountered in the *limb* regions. On the *terminator* – i.e., at the time of sunrise (or sunset), when the peak P just intercepts the first (or last) rays of the rising (or setting) Sun,

$$s_{\max} = (a+h)\sin\omega\tag{5.33}$$

and, from (5.2) and (5.3) it follows that

$$v = \omega,\tag{5.34}$$

such that

$$\cos v = a/(a+h) = \cos\omega,\tag{5.35}$$

where ω stands then for the selenocentric arc at which P becomes visible beyond the terminator in the direction of incident sunlight. This angle is difficult to measure directly; but can be computed with the aid of the equality $\omega = v$ from known values of the position β_P, λ_P of the peak in question, as well as of the selenocentric position A, D of the Sun at the time of observation. This method was, in fact, first used to estimate the altitudes of the lunar mountains by Galileo Galilei. We may wish to add only that, in such a case, h need not stand for the height of any isolated peak, but also for that of any plateau along the terminator. If the Moon were a perfectly smooth sphere, the sunrise (or sunset) terminator would be an ellipse. Since the terminator is defined as the locus of points at which the Sun just rises above (or sets below) the lunar horizon, any irregularities of its outline or departures from an ellipse would be indicative of the differences in level at different latitudes intersected by it.

Throughout all our discussion of the geometry of the shadow method for the determination of lunar relative altitudes we have, in fact, tacitly assumed so far that the shadow of the peak P is cast on a sphere. The lunar surface is, however, by no means smooth in detail; and the shadow method – furnishing as it does information on altitude difference between points P and S – should enable us to ascertain not only the relative altitude of any particular eminence above the surrounding landscape, but also, in principle, any irregularities or undulation of ground on which the shadows are cast. In order to outline the steps by which this can be done, let us return to Figure 5.1, and suppose that the actual distance from the Moon's center to the point S is, not a, but $a + \delta a$, with δa denoting a local deviation of the mean Moon level (such as a difference in level between the foot of a mountain and the tip of its shadow). If so, Equation (5.1) continues to hold good irrespective of surface irregularities; but Equations (5.2) and (5.3) should be replaced by

$$(a + \delta a) \sin \omega = s \cos v \tag{5.36}$$

and

$$(a + \delta a + h) \cos v = (a + \delta a) \cos(v - \omega), \tag{5.37}$$

leading to

$$\frac{h}{a + \delta a} = \frac{\cos(v - \omega)}{\cos v} - 1 \tag{5.38}$$

in place of (5.4), where

$$\sin \omega = \frac{r'_s \sin \mu \cos v}{(a + \delta a) \sin(\theta + \mu)} \tag{5.39}$$

exactly; with $a + \delta a$ replacing a in Equation (5.6) for $\mathbf{r}'_s$; but the definitions of θ and v unchanged.

Should the shadow of P be cast on a sphere, the measured angular shadow length μ should be a smooth function of the time varying as the Sun rises or sets on the Moon in a manner predictable from our theory. If, however, the landscape on which the shadow is cast is uneven, a plot of μ versus the time should show irregularities, depending on the local variation of δa. Suppose that we draw a smooth curve by free hand through such irregularities, which we shall use for a definition of the local mean Moon level. Deviations $\delta\mu$ from such a curve can then be translated into the corresponding undulations δa of ground on which the shadows are cast with the aid of the formulae given in the preceding paragraph, in which angles θ and v continue to be given by Equations (5.14) and (5.15) or (5.30).

In conclusion of the present discussion of the geometry of shadows cast by sunlit mountains, it should be stressed that we have so far assumed – for the sake of simplicity – the Sun to act as an illuminating point-source of light. In reality, of course, the apparent angular diameter of the Sun as seen from the Moon amounts to very approximately $\frac{1}{2}°$; and this fact alone is bound to provide all lunar shadows with a *penumbra* even in the complete absence of any atmosphere – covering regions a part of the apparent solar disc would be set for the observer on the ground. The intensity of illumination at any point of this penumbral zone should depend on the brightness of the visible segment of the Sun; and this will vary from full light to complete darkness over a strip whose width should depend on the altitude of the rising (or setting) Sun – becoming the greater, the lower the shadow-casting obstacle.

For this reason, dependable information on the angular length of lunar shadows cannot easily be obtained by any visual settings possible with a micrometer, but must be sought by micro-densitometric tracings – a method introduced in lunar studies and developed to its present state by the Manchester group of astronomers led by Zdeněk Kopal (for its partial summary cf., e.g., Kopal *et al.*, 1961). Illustrative examples of different variants of this technique are shown on Figures 5.3, 5.4; and its applications, on Figures 5.5 (Theophilus) and 5.6 (Archimedes); while the results obtained from an analysis of this latter formation are diagramatically exhibited on Figure 5.7.

In order to make proper use of such microdensitometric records – and, in particular, to determine the point, in the penumbral band, at which the center of the apparent solar disc (whose coordinates occur in Equations (5.14) and (5.15) just clears the lunar horizon – it is, however, necessary first to investigate the expected distribution of light within the penumbral zone; and this can be approached in the following manner.

Let, as on Figure 5.1, the peak P cast a shadow in the light of a light source, having the position of the Sun and an angular semi-diameter of $\varrho_\odot = 15'59''.6$. The width $f_{1,2}$ of the penumbral zones on either side of S will clearly be given by

$$f_{1,2} = \frac{s \sin\varrho_\odot}{\sin(v \pm \varrho_\odot)} \tag{5.40}$$

in the SP'-direction; and their projections $f_{1,2}$ on a plane tangent to the lunar surface at S become

$$f'_{1,2} = \frac{s \sin\varrho_\odot}{\sin(\varrho_\odot - \omega \pm v)}; \tag{5.41}$$

Fig. 5.3.   Miss Ellen B. Finlay of the University of Manchester, sorting out lunar film negatives for micro-
densitometric measurements illustrated on Figures 5.4–5.7.

the upper and lower sign in the denominators referring to the parts of the penumbra
interior and exterior to S, respectively. The total width of the penumbral zone then
becomes equal to

$$f_1' + f_2' = \frac{2 \sin \varrho_\odot \, \cos v \, \sin(\varrho_\odot - \omega)}{\sin(\varrho_\odot - \omega + v) \, \sin(\varrho_\odot - \omega - v)},$$ (5.42)

where the angle v continues to be given by (5.15), while s and ω follow from (5.1) and
(5.5).

Fig. 5.4.   Dr Thomas W. Rackham at work with a microdensitometer with which thousands of photo-metric transcripts of lunar craters of the type shown on Figure 5.5 have been obtained at the University of Manchester between 1958–1968.

Let, moreover, the position of any point within the penumbral zone of the lunar surface in the direction of incident sunlight be characterized by a single coordinate x measured positively outwards from S and normalized so as to assume the values ± 1 at the ends of the penumbra. If so, and if the lunar horizon acts like a straight occulting edge (or, which is more likely, the horizontal irregularities are small in comparison with the solar semi-diameter), then it can be shown (cf., e.g., Kopal, 1959a, p. 207) that the fractional illumination $I(x)$ at any point of the penumbra should vary as

$$\pi I^{U}(x) = \pi - \cos^{-1} x + x \sqrt{(1 - x^2)} \qquad (5.43)$$

if the apparent solar disc were uniformly bright, and

$$4 I^{D}(x) = (1 + x)^2 \, (2 - x) \qquad (5.44)$$

if it were completely darkened at limb.

   In actual fact, the solar disc is known to be partially darkened at the limb; and in the yellow light ($\lambda = 0.56 \, \mu$) is coefficient of darkening u is known to be approximately equal to two-thirds; and if so, we should expect (cf., again Kopal, 1959a, p. 308) that

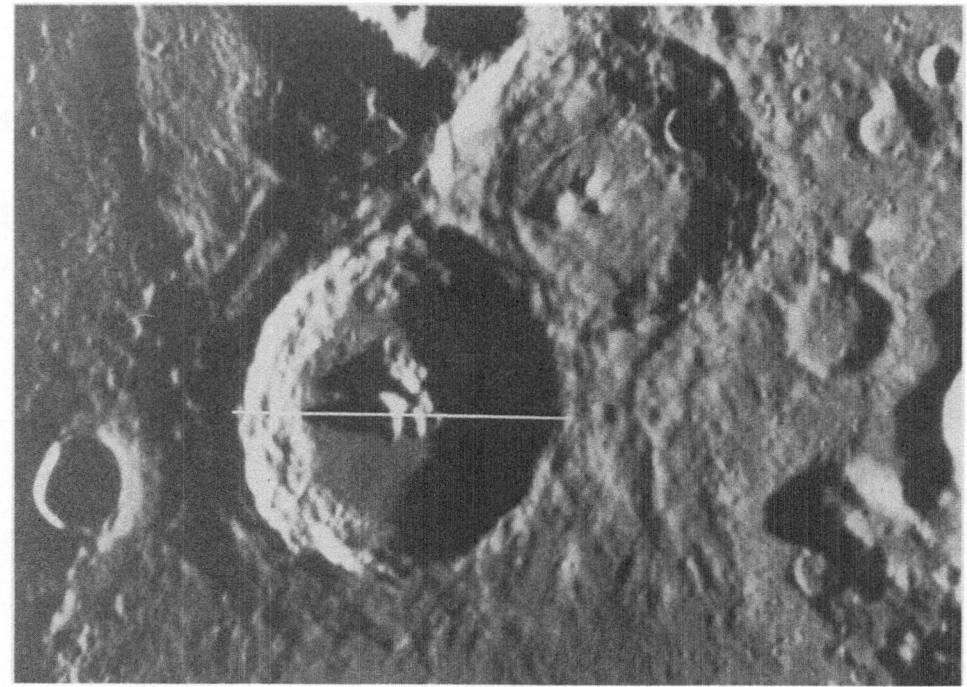

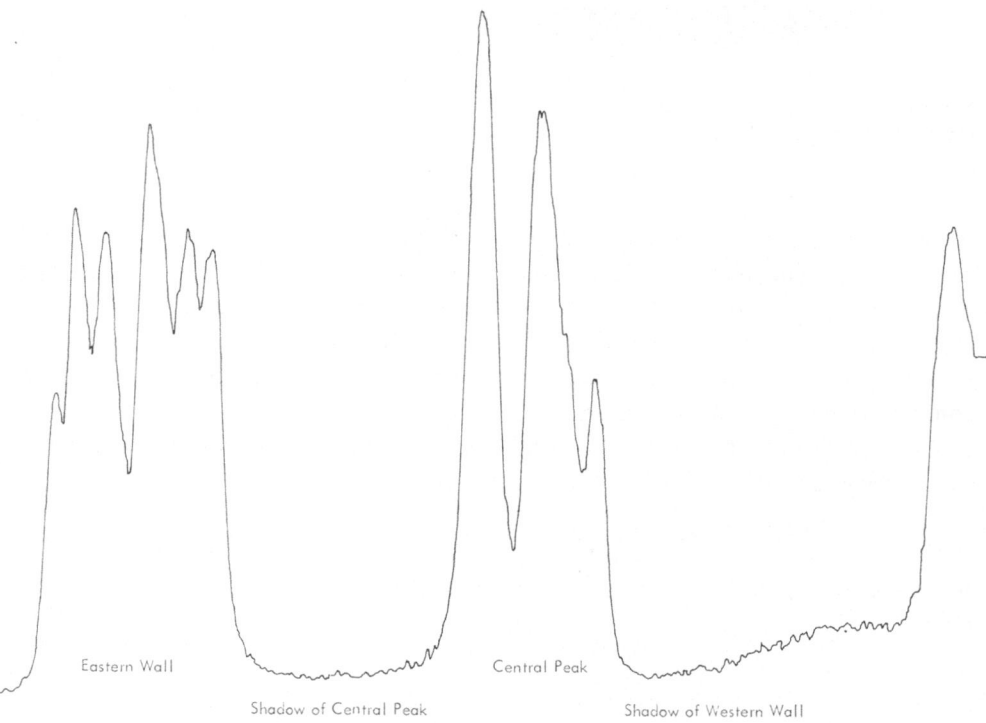

Eastern Wall

Shadow of Central Peak

Central Peak

Shadow of Western Wall

Fig. 5.5. Sunset over the craters Theophilus and Cyrillus, photographed with the 24-in. refractor of the Observatoire du Pic du Midi (Manchester Lunar Programme). The white line across Theophilus indicates the direction of the microdensitometric tracing reproduced below.

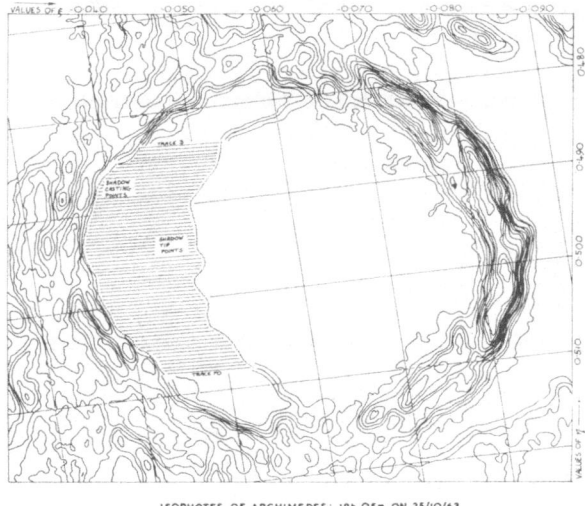

ISOPHOTES OF ARCHIMEDES: 19h 05m ON 25/10/63

CONTOUR VALUES REFER TO THE RECORDED PLATE DENSITY, THE SUPERIMPOSED LINES

ARE THOSE OF THE STANDARD ORTHOGRAPHIC MAP GRID AT INTERVALS OF 0.01 OF

A LUNAR RADIUS

Fig. 5.6. Direct photograph of sunrise over the crater Archimedes in the Eastern part of Mare Imbrium, taken with the 24-in. refractor of the Observatoire du Pic du Midi on 25 October, 1963 (above), and its microdensitometric transcript (below) used to study the vertical relief of this formation by the shadow method (after Jones, 1965).

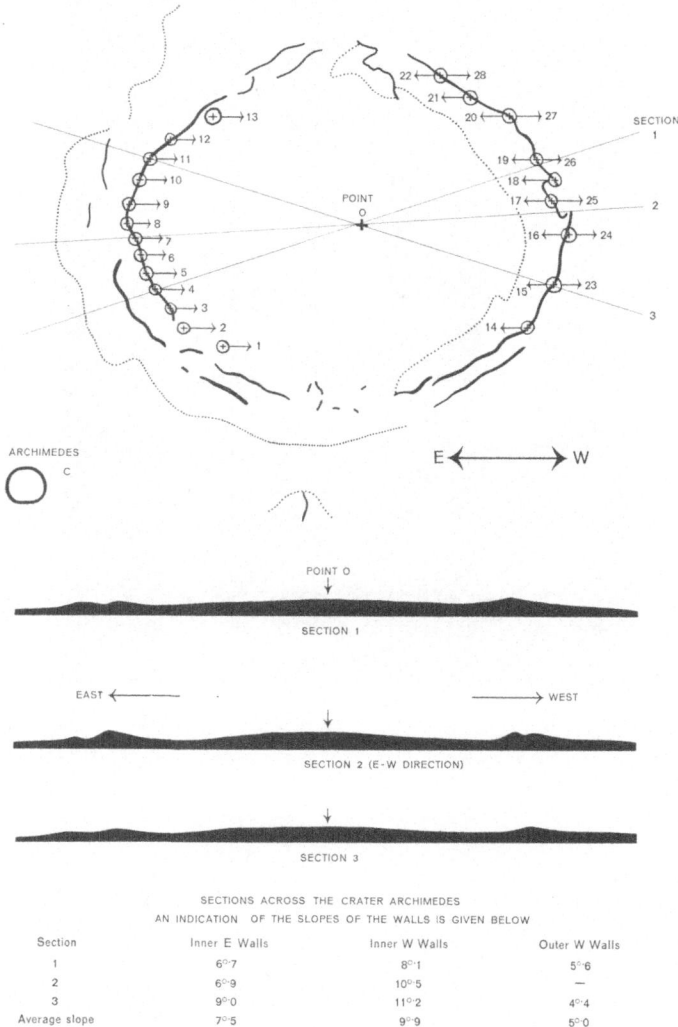

Section	Inner E Walls	Inner W Walls	Outer W Walls
1	$6°.7$	$8°.1$	$5°.6$
2	$6°.9$	$10°.5$	—
3	$9°.0$	$11°.2$	$4°.4$
Average slope	$7°.5$	$9°.9$	$5°.0$

Fig. 5.7. Topography of the lunar crater Archimedes (after Turner, 1959).

$$I(x) = \frac{3(1-u)}{3-u} I^{U}(x) + \frac{2u}{3-u} I^{D}(x) =$$
$$= \tfrac{1}{7}\{3I^{U}(x) + 4I^{D}(x)\}. \tag{5.45}$$

The essential characteristic of this expected distribution of light is the fact that both functions $I^{U}(x)$ and $I^{D}(x)$ as defined by Equations (5.43) and (5.44) – and, therefore, their weighted mean (5.45) – possess an inflection point at $x=0$.* The variation of intensity in the penumbral shadow turns out, therefore, to be most rapid when the *center* of the apparent solar disc just rises above the lunar horizon at S. This fact

---

* This would, moreover, continue to be true for *any* distribution of brightness over the apparent solar disc – provided that it retains radial symmetry.

enables us then to *define* the position of *S* as the point at which a micro-densitometric tracing (on the intensity scale) traversing the penumbral zone exhibits inflection.

What would happen if the shadow-casting obstacle on the Moon were so low (like the lunar wrinkle ridges, for instance) that the entire width of the visible shadow were, in fact, a penumbra? The measurement of the shadow lengths μ would then of course cease to possess any meaning. However, any unevenness of even gently undulating ground can, in principle, be established from the photometric measurements of surface brightness which, at low angles of incidence, should vary as the cosine of the angle between incident sunlight and the surface normal (in accordance with Lambert's law): as the former direction is a known function of the time, photometric measures can be used to determine the latter. Such a method was indeed worked out by Van Diggelen (1951) and applied by him to a determination of the profiles of lunar wrinkle-ridges in the region of the Mare Imbrium, with encouraging results.

The efficiency of the shadow method for determination of the relative altitudes of lunar mountains is at its best when the line of sunrise or sunset terminator (along which the length of the shadows magnifies the altitude difference to maximum extent) passes through the center of the apparent lunar disc (i.e., the domain of minimum foreshortening). At the limb of the Moon, the limits of precision with which one can determine the altitude of lunar mountains is given by the resolving power of the telescope employed, which for a 24-in. (60 cm) aperture in yellow light is (by Rayleigh's criterion) close to 500 m. However, near the center of the apparent lunar disc, the vertical heights are greatly magnified in the shadows cast in the oblique rays of the Sun. Simple geometry reveals that, at a time when the whole disc of the Sun just appears above the lunar horizon, the ratio of the length of the shadow to the height of the object casting it is equal to $\cot 0°5 = 115$ (and exceeds 115 in the penumbral zone, illuminated by only a part of the solar disc); though when the Sun's center has risen to 2° above the lunar horizon, this magnification ratio has diminished to 27. If, therefore, photographs can identify on the Moon at sunrise the shadows (say) 1 km in length, these should be indicative of the presence of vertical obstacles less than 10 m in height; and of even smaller obstacles in the penumbral zone (which is, of course, not more than approximately 18 km wide on the Moon at any time).

Have such objects been actually seen, or photographed, on the Moon from the distance of the Earth? Unfortunately, virtually all photographs obtained so far are ill-suited to be used for an appropriate search, as their exposure times chosen so as to bring out the entire field covered by each plate were invariably much too short to reveal any details inside the relatively narrow twilight strip. It may perhaps come to the reader as a surprise that at most – if not all – lunar photographs he may have seen, the apparent terminator marked the region where the Sun has risen already several degrees above the lunar horizon; for exposures capable of revealing details in the zone illuminated by a lower Sun would have left the rest of the field very largely burnt out. Moreover, direct visual observations would likewise be of little avail, as the twilight (penumbral) zone would appear too dark to the eye to discern any contrast.

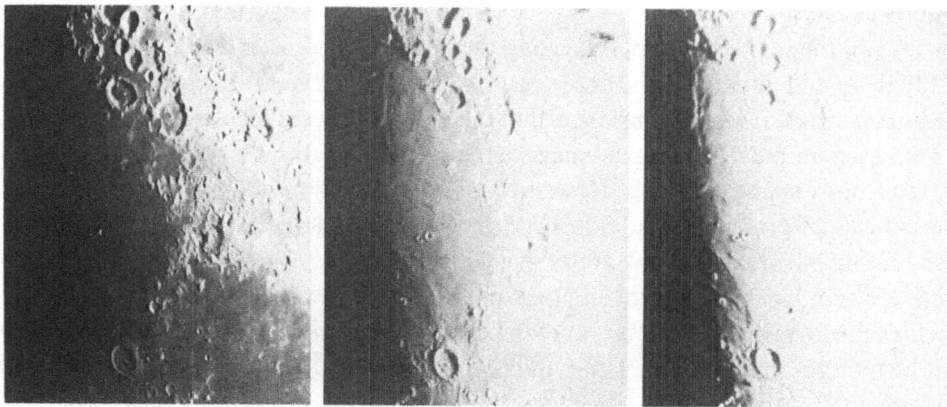

Fig. 5.8. Sunset over Mare Foecunditatis, taken in the $f/15$ focus of the 43-in. reflector at Pic du Midi on March 2, 1964. Exposures: left, 0.025 s; center, 0.5 s; right, 2 s – all within the same minute. Note increased definition of the terminator with increasing exposure time.

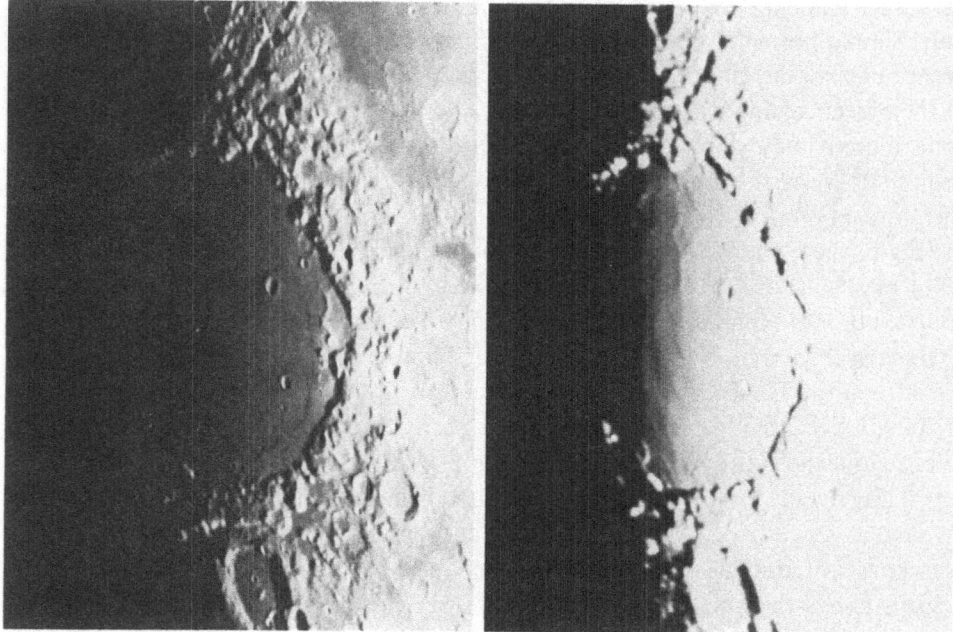

Fig. 5.9. Sunset over Mare Crisium, photographed on low-contrast plates with the 24-in. refractor at Pic du Midi on January 2, 1964. Exposures: left, 1 s; right 50 s – both within the same minute. Note increasing definition of the terminator with the exposure.

In order to illustrate this point, we reproduce on Figures 5.8 and 5.9 two pairs of photographs of the same terminator region of the Moon and obtained practically at the same time – one with a 'normal' exposure, the other 'overexposed' sufficiently to bring out details in the actual penumbral zone. The reader can judge for himself the extent to which these overexposures resulted in an increase of information in the twilight zone, much of which has not been known before. Indeed, it is quite probable that 'overexposed' lunar photographs, as reproduced on Figures 5.8 and 5.9, can enable us to bridge – at least in certain parts of the apparent lunar disc – to some extent the gap between the Earth-bound and space-borne cameras recording the surface of our satellite.

In conclusion, one last point – though a hardly surprising one – should be stressed: namely, the limitation of all shadow topographic work arising from the fact that the shadow-casting sunlight illuminates the Moon always from very much the same direction. In point of fact, the illuminating sunlight can vary in direction only as much as permitted by the libration of our satellite in latitude; and this is a very small amount. In consequence, the magnification of small altitude differences by the shadow effect will facilitate detection of those surface irregularities which are situated normally to the incident sunlight, and discriminate against objects that are parallel with it. For instance, a rille or wrinkle ridge in the mare plains will be more easily detected if it runs parallel with the meridian if it were in the direction of a latitude circle. This observational selection forced upon us by the almost uni-directional nature of the illuminating light source may exert serious effects (the majority of known wrinkle ridges do indeed run more or less in the direction of the meridians), and must be kept in mind in any work on the geological interpretation of the lunar landscape.

Throughout this chapter we have also been concerned mainly with a determination of *vertical* differences on the lunar surface, which offers many points of methodological interest. The determinations of *lateral* dimensions of lunar formations are much more straightforward – from photographs or at the telescope; for a conversion of angular measurements into absolute units requires merely a knowledge of the instantaneous distance of the feature in question on the lunar surface from the observer. For instance, a circular crater on the surface of the Moon will, from the terrestrial vantage point, generally appear in projection as an ellipse of semi-major axis a' along an arc parallel with the limb, and semi-minor axis $b' = a' \cos k$, where k denotes the selenocentric angle of the respective formation.

Let, moreover, s' denote the apparent topocentric semi-diameter of the Moon at the time of observation (related with the geocentric semi-diameter s by Equation (2.19). If so, the absolute value of the true semi-major axis a will be given by the equation

$$a = 1738 (a'/s') (r'_s/r') \text{ km}, \tag{5.46}$$

where r'_c denotes the topocentric distance of the respective lunar formation; and r', the topocentric distance of the Moon's center. As, however, from (5.6) it follows that

$$r'_s/r' = 1 - \sin s' \cos \eta + \cdots \tag{5.47}$$

and, very approximately, $\eta \approx \kappa$, Equation (5.46) can to a sufficient precision be re-written as

$$a = 1738\,(a'/s')\,(1 - \cos\kappa\,\sin s')\ \text{km} \tag{5.48}$$

and

$$b = 1738\,(b'/s')\,(1 - \cos\kappa\,\sin s')\,\sec\kappa\ \text{km}. \tag{5.49}$$

These equations specify the extent to which the time dimensions of a lunar crater will be vitiated in projection, on account of the fact that all parts of their image are not at the same distance from the observer.

Methods for the determination of relative elevations on the Moon from the measurements of illumination and shadows cast in the oblique rays of the rising or setting Sun possess a distinguished pedigree going back to the earliest days of modern astronomy. A detection of vertical inequalities on the surface of our satellite belonged among the first telescopic discoveries of Galileo Galilei (1610), who was also the first to attempt estimates of the altitudes of lunar mountains from the angular distance beyond the terminator at which they become sunlit in the rays of the rising Sun. Needless to stress, Galileo – lacking any kind of even a rudimentary micrometer – was in no position to perform actual measurements with his cannocchiale; and the altitudes assigned by him to some (unidentified) peaks – rendering them higher than Mount Everest in the terrestrial Himalayas – represented gross overestimates of the actual facts, as was pointed out only a little later by Hevelius (1647).

While Galileo Galilei can, therefore, be regarded as the father of observational studies of lunar topography, the theoretical basis for the interpretation of such observations was laid down by no one lesser than Johannes Kepler – the second 'founding father' of modern astronomy in the first half of the 17th century. In his *Somnium* (1634) – and, more specifically, in its Appendix in the form of a *Letter to Father Paul Guldin* – Kepler outlined quite clearly the geometrical basis of lunar hypsometry expounded in this chapter.

The first investigator actually to measure the altitudes of lunar peaks from the extent of their visibility beyond the terminator was William Herschel (1787), using a micrometer at his 6-ft telescope magnifying 222 times. Although Herschel was wont to exaggerate the precision of his measurements (listing them to $0\overset{''}{.}001$, while their actual errors were several hundred times as large), he was correct in his realization that lunar mountains are, in general, much lower than was thought by Galileo or even Hevelius; the majority of them being, according to Herschel, between $\frac{1}{2}$ and $1\frac{1}{2}$ miles in height.

Herschel's work was soon followed by Schröter (1791, 1802); and, in the 19th century, by Beer and Mädler (1837) and Schmidt (1878) who provided the bulk of our knowledge of the altitude of lunar mountains available up to the middle of the 20th century. Beer and Mädler, together with Schmidt, abandoned the Galilei-Herschel method of timing the beyond-the-terminator appearance of sunlit peaks for the measurements of the shadows cast by them on the surrounding landscape; and in more modern days this method was adapted for the use of photographic cine-technique by McMath *et al.* (1937).

The geometrical basis underlying the shadow method had been credited to Olbers (see, e.g., Graff, 1901); and has since been exhaustively elaborated by Kopal and his collaborators at the University of Manchester in England, working at the Observatoire du Pic du Midi in the French Pyrénées (2864 m above sea-level) as a part of the collaborative effort supported by the U.S. Air Force. This work has been partly described in a series of publications that have appeared since 1960 (cf., e.g., Kopal, 1959, 1960, 1961a, b, 1962; Kopal and Rackham, 1962; Sudbury, 1965; Jones, 1965) who have, by a joint effort, brought its techniques to the level at which it has been

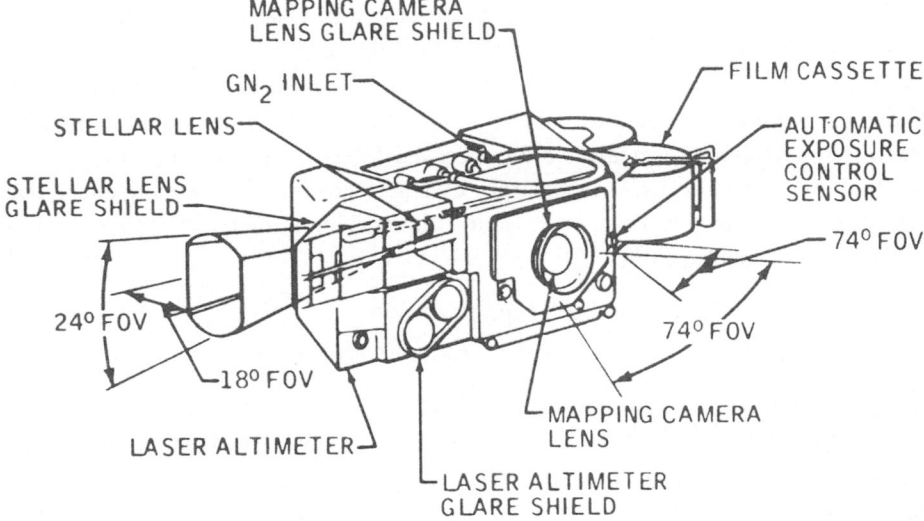

Fig. 5.10.   Apollo 15–17 photographic cameras and laser altimeter.

presented in this chapter. This is, in particular, true of the treatment of penumbral phenomena on the Moon (Kopal, 1962); of the method of reduction of the shadow measurements from photographs taken aboard the lunar spacecraft (Kopal, 1966); or of the use of 'overexposed' photography at the terminator (cf. Kopal, 1967; Rackham, 1967; Sudbury, 1967).

The methods developed and used by these investigators have provided a large majority of the hypsometric data incorporated in lunar maps of the last ten years, and described in the second part of this book.

The determination of elevation and positional differences within a local area through spacecraft photography, is accomplished by employment of the same principals of photogrammetric intersection as described in Chapter 3, on Selenographic Coordinates. However, more precise relative heights and positions are obtainable as systematic errors in camera station positions and orientations have minimal effect on determination of coordinate differences in local solutions. The principal factors affecting the precision of relative differences are the scale, resolution and geometric fidelity of the photography, with accuracy of height determinations being particularly dependent on favourable (large) intersection angles. That is, increased height accuracy is a function of the ratio

$$\frac{\text{distance between camera stations}}{\text{distance from camera station to lunar surface}}$$

These principals were applied to the determination of the detailed topographic configuration of potential Apollo landing sites with Lunar Orbiter Mission photographs. Photogrammetric solutions covering selected site areas were constructed which employed photographic coverage acquired from different orbital revolutions (increased distance between camera stations) and minimal orbital altitudes, enabling derivation of local differences in heights of features, to within a few meters in optimum cases.

The most extensive facility for determining precise relative lunar coordinates is provided by combined usage of panoramic and frame photography acquired by Apollo missions 15–17. The long focal length (24-in.), high resolution panoramic camera provides relatively large scale photography portraying minute lunar detail and advantageous intersection angles. Its principal drawback is limited geometric stability caused by the dynamic nature of the camera. In practice, the panoramic photogrammetric intersection solution is constrained to an array of points developed from overlapping mapping camera (see Figure 5.10) photographs, enabling determination of relative coordinate differences to a 5–10 m accuracy.

References

Beer, W. and Mädler, J. H.: 1837, *Der Mond*, Simon Schropp, Berlin.
Diggelen, J. van: 1951, *Bull. Astron. Inst. Netherlands* **11**, 283.
Fujinami, S.: 1952, *Publ. Astron. Soc. Japan* **4**, 115.
Fujinami, S., Ina, T., and Kawai, S.: 1954, *Publ. Astron. Soc. Japan* **6**, 67.
Galilei, Galileo: 1610, *Nuncius Sidereus*, Baglioni, Venice (English translation by E. S. Carlos, published by Dawsons of Pall Mall, London, year not given).

Hayn, F.: 1914a, *Abh. König. Sächs. Gesell. Wiss.* **32**, 1–113.
Hayn, F.: 1914b, *Astron. Nachr.* **199**, 261.
Herschel, W.: 1787, *Phil. Trans. Roy. Soc.* **77**, 229.
Jones, M. T.: 1965, M.Sc. Thesis, Univ. of Manchester.
Kopal, Z.: 1959, *Amer. Scientist* **47**, 505.
Kopal, Z.: 1960, in *Proc. 1st COSPAR Space Sci. Symp. Nice* (ed. by H. K. Kallmann-Bijl), Amsterdam, pp. 1123–1131.
Kopal, Z.: 1961a, 'Studies in Lunar Topography', Air Force Cambridge Res. Lab., GRD Res. Note No. 67, p. 188.
Kopal, Z.: 1961b, *Rev. Internat. Council Sci. Unions* **3**, 173.
Kopal, Z.: 1962, in *Physics and Astronomy of the Moon* (ed. by Z. Kopal), Academic Press, New York and London, pp. 231–282.
Kopal, Z.: 1966, *An Introduction to the Study of the Moon*, D. Reidel Publ. Co., Dordrecht; Chapter 14, pp. 187–206.
Kopal, Z.: 1967, in *Measure of the Moon* (ed. by Z. Kopal and C. L. Goudas), D. Reidel Publ. Co., Dordrecht, pp. 407–413.
Kopal, Z. and Rackham, T. W.: 1962, in Z. Kopal and Z. K. Mikhailov (eds.), 'The Moon', *IAU Symp.* **14**, Academic Press, New York and London, pp. 343–360.
Kristenson, H.: 1954, *Arkiv Astron.* **1**, 411.
McMath, R. R., Petrie, R. M., and Sawyer, H. E.: 1937, *Publ. Univ. Obs. Michigan* **6**, 67.
Nefediev, A. A.: 1958, *Bull. Engelhardt Obs. Kazan*, No. 30.
Rackham, T. W.: 1967, in *Measure of the Moon* (ed. by Z. Kopal and C. L. Goudas), D. Reidel Publ. Co., Dordrecht, pp. 414–423.
Schmidt, J. F. J.: 1878, *Die Charte der Gebirge des Mondes*, Ergänzungsband, Dietrich Reimer, Berlin.
Schröter, J. H.: 1791, *Selenotopographische Fragmente* I, Lilienthal.
Schröter, J. H.: 1802, *Selenotopographische Fragmente* II, Göttingen.
Sudbury, P. V.: 1965, M.Sc. Thesis, Univ. of Manchester.
Sudbury, P. V.: 1967, in *Measure of the Moon* (ed. by Z. Kopal and C. L. Goudas), D. Reidel Publ. Co., Dordrecht, pp. 424–432.
Turner, G.: 1959, *Astron. Contr. Univ. Manchester, Ser. III*, Nos. 72 and 73.
Watts, C. B.: 1963, *Astron. Papers Amer. Ephemer. and Nautical Almanac*, Vol. 17.
Weimer, Th.: 1952, *Atlas des profils lunaires*, Publ. de l'Observ. de Paris.
Whitwell, T.: 1929, *J. British Astron. Assoc.* **39**, 255.

CHAPTER 6

U.S. AIR FORCE LUNAR MAPPING

Prior to the establishment of the National Aeronautics and Space Administration (NASA), the United States Air Force (USAF) foresaw the need for acquiring improved technical data on lunar surface features. To this end the Air Force Cambridge Research Laboratory (AFCRL) instituted several lunar projects.

The first of these projects involved an AFCRL contract in the fall of 1957 with Gerard Kuiper of Yerkes Observatory to assemble a collection of the best available photographs of the Moon, which ultimately led to the publication of the *USAF Lunar Atlas*. Over 1200 prints were evaluated, from which 281 photographs presented in the main body of the Atlas were selected. These represented the best materials available at Mt. Wilson, Lick, McDonald, Yerkes, and Pic du Midi Observatories. This Atlas was the culmination of efforts by many different groups, in widely separated areas, bringing together the most significant photographic coverage of the Moon (as of 1958) to be assembled under one cover.

Completion of the *USAF Lunar Atlas* was a cooperative endeavor between AFCRL and USAF Aeronautical Chart and Information Center (ACIC).* The reproduction was accomplished by a commercial printer but the final assembly and distribution was made by ACIC in March, 1960. A commercial edition of this atlas, known as the 'Kuiper Atlas', was made available to the general public through the University of Chicago Press.

Almost concurrent wfth the Lunar Atlas project, another was instituted at the University of Manchester, England. Determining heights of lunar mountains by measuring the length of the shadow cast by the peak dates back to the 18th century. But it remained for a small group of astronomers at the University of Manchester, working under the direction of Prof. Zdeněk Kopal, to update and refine this technique. This work was performed under an AFCRL contract issued in 1958. They observed 35 mm time-lapse photographs taken at Pic du Midi Observatory of lunar sunrise and sunset over specific parts of the Moon. To accomplish this, they modified a 35-mm movie camera so that it would take exposures at the rate of 3 per minute, then mounted it on the 24-in. refractor with an optical bench attached to the lower end of the telescope. By this method they acquired some 12 000 individual exposures of selected areas of the lunar surface during a two-year period. The research phase of the Manchester contract was completed during the summer of 1960.

Through the lunar research projects at AFCRL, ACIC became involved with the

---

* Effective July 2, 1972, ACIC became the Defense Mapping Agency Aerospace Center (DMAAC).

Moon in the fall of 1959. NASA, a government agency being organized at that time, was making plans for a manned Earth-orbital flight to be named Mercury, and discussions were under way for possible manned journies to the Moon. Thus, formative plans began at ACIC to produce maps to support the planned Mercury flight and to map the Moon. The ensuing programs at ACIC satisfied two objectives: general purpose lunar mapping, and space support mapping. (The general purpose lunar maps will be covered in this chapter and the USAF space support mapping in a subsequent chapter).

1. Lunar Mosaic

The first lunar cartographic item published by ACIC was a photographic mosaic of the Moon, identified as the *USAF Lunar Reference Mosaic* LEM-1, (Lunar Earthside Mosaic) 1:5 000 000 (lunar diam 27 in.). It was compiled under the supervision of ACIC cartographer Howard Holmes.

The USAF Lunar Reference Mosaic, LEM-1, shown in Figure 6.1, is a composite

Fig. 6.1. USAF Lunar Reference Mosaic (LEM-1).

photomosaic of the Moon produced from selected photographs taken at Mt. Wilson, McDonald and Pic du Midi Observatories. Sections of twenty-four photographs were chosen which provided a constant Sun angle to maintain a uniform portrayal of lunar craters and prominences.

This mosaic was compiled on an orthographic projection which shows the Moon as a sphere as viewed from an infinite distance. Position was determined from the seleneographic control established primarily from the measures of Franz and Saunder as listed in the *Orthographic Atlas of the Moon* compiled by David Arthur and Ewen Whitaker in 1961. Each photograph was copied to a common scale and rectified to mean libration in order to match or fit adjacent sections.

The *USAF Lunar Reference Mosaic* was originally compiled in February 1960 and published at two scales, 1:5000000 (LEM-1) and 1:10000000 (LEM-1A). In 1962, LEM-1 was recompiled from improved photo imagery and issued at scales of 1:10000000; 1:5000000 and 1:2500000, the latter in two sheets and titled *USAF Lunar Wall Mosaic* (LEM-1B).

This series of lunar mosaics has been one of the most popular items in the ACIC lunar map inventory. The mosaics have been very much in demand as a wall display and as a base for various indices.

All three sizes of LEM are lithographed in duotone blue and gray against a solid black background.

2. Lunar Astronautical Chart (LAC)

During the same time as the USAF Lunar Mosaic (LEM-1) was in work, ACIC began formulating plans for charting the moon under the project direction of ACIC staff cartographer Robert Carder. The question might be asked, why produce a chart of the Moon when photographic coverage was already available? The answer: a well prepared chart can reflect the best qualities of innumerable photographs, can remove the apparent distortion of the photo image, and can be supplemented by telescopic observations in which the resolution is two to three times better than the average lunar photograph. By applying cartographic techniques, the end product results in improved control of features, true shapes, and accurate relative placement of features.

ACIC selected the scale of 1:1000000 (16 miles to the inch) for charting the Moon because this scale seems to be compatible with the maximum resolution obtainable from lunar photographs, supplemented by visual telescopic observations and because the Air Force for years had maintained a 1:1000000 scale coordinated series of charts covering the Earth, the World Aeronautical Chart (WAC) series. Thus, the 1:1000000 scale Lunar Astronautical Chart (LAC) was born.

Work on the LAC series commenced in October 1959 under a task force headed by ACIC cartographer Howard Holmes and assisted by ACIC cartographers Jerry Higgins and Charles Moore. The entire Moon was evenly divided into 144 areas, each 22″ × 29″. Each chart was given the name of the most prominent feature in the area, for example, Kepler LAC-57, Copernicus LAC-58, Mare Vaporum LAC-59, etc.

The numbering system started with 1 at the North Pole, then followed consecutively to the right, around the lunar sphere, ending with sheet 144 at the South Pole.

The 144 LAC sheets are laid out on three types of projections. The Mercator covers 16° N-S lat. embracing two bands of charts which join perfectly to form a continuous strip around the lunar equator. The Lambert conformal representing four bands of charts covers the area from 16° to 80° N-S latitude, and the polar stereographic for the two polar charts, which were never produced. The work in laying out the LAC series was accomplished by ACIC cartographer William Cannell.

In coming to grips with the actual task of making a lunar chart, ACIC cartographers were not concerned with the problems of portraying a mass of cultural data such as shown on Earth charts; their main concern was depicting the lunar landscape and determining elevations of these features.

ACIC cartographers concluded that lunar features could be artistically drawn and airbrushed with India ink on translucent plastic sheets. Thus, the drawings show craters, hills, domes, pressure ridges, etc., as though they were lighted from an afternoon Sun which varies so that the angle of illumination nearly matches the slope angle of the features. In this way, there are no cast shadows to cover up detail, yet the very low or shallow features receive sufficient shading to be clearly discernible.

Another important surface characteristic is the ray system – distinctive features evident in full-moon photographs. Though it is unnatural to see the relief and rays on a single lunar photograph, both features are shown on the same chart by printing them in two colors: olive-green for relief and blue-gray for the full-moon ray system. These colors were selected for their aesthetic value because the true color of the moon was not known in 1960.

Contours printed in brown to complement the olive-green relief are related to an assumed spherical figure of the Moon whose radius is established at 1738 km. The contour interval of 300 m was selected for its convenience in converting to the American usage of the 1000-ft interval. So that most of the contours would have positive values, ACIC established a zero point or 'Moon datum' at 2.6 km less than the 1738 km radius.

ACIC published the first LAC (LAC-58), which is shown in Figure 6.2, in Feb. 1960. It was compiled on a Lambert conformal conic projection from 0° to 14° N lat. Subsequently, it was recompiled on a Mercator projection from 0° to 16° N lat. The first LAC's were compiled entirely from existing photographs without benefit of visual telescopic observations. Also they were drawn with North at the top to conform with terrestrial mapping. This was contrary to many of the 18th and 19th century lunar maps which were drawn with South at the top to conform with the inverted image of the Moon as viewed in telescopes in the northern hemisphere. In 1960, the International Astronomical Union (IAU) General Assembly (Berkeley) acknowledged the need to define cardinal direction for lunar mapping and adopted the following resolution: 'Astronautical maps for direct exploration purposes are to be printed in agreement with ordinary terrestrial mapping. North being up, East at right and West at left'.

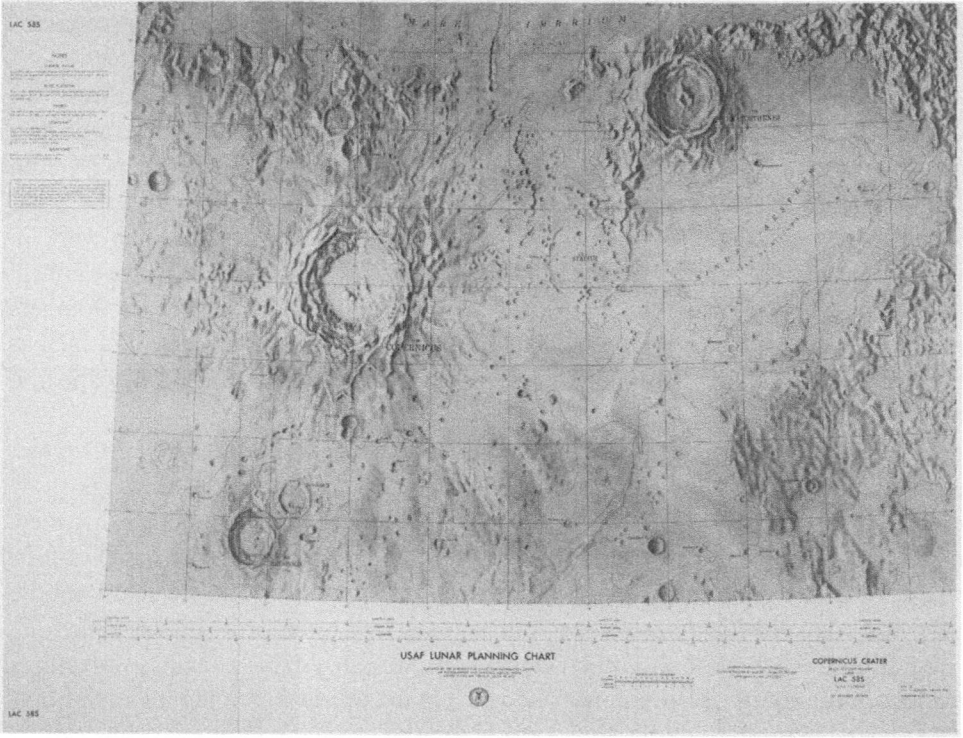

Fig. 6.2.   The first LAC chart of the Moon, LAC-58 published by ACIC in Feb. 1960. This chart was
drawn from photographs without benefit of telescopic observations.

Early in 1960, ACIC recognized that a good telescope, located where the best
viewing conditions would prevail was most vital to the success of a modern lunar-
mapping program. A search for such a telescope led ACIC to Lowell Observatory
and the availability of their 24-in. refractor. Arrangements were made for part-time
use of this telescope and eventually in September 1961 a permanent ACIC observation
unit was established at Lowell. (The ACIC work accomplished at Lowell Observatory
from 1961 through 1969 is being covered in the following chapter; but shown in
Figure 6.3 is LAC-58 based on visual observations at Lowell.)

ACIC determined relative lunar heights on the LACs by the shadow-measuring
technique as indicated in Figure 6.4. This technique involves the treatment of the
shadow measurement, ephemeris data and time of exposure. After these factors have
been determined, the length of the shadow in the photograph is converted to the
physical length of the shadow on the lunar surface. Knowing the physical length of
the shadow and the Sun angle at the time of exposure, one can determine the height
of the feature relative to the surrounding area. To establish a network of 300-m
contours for a LAC area, several thousand of these measurements were needed for
each chart.

Many photographs of varying Sun angles were required to accomplish an effective

Fig. 6.3.   LAC-58, 2nd edition, April 1964, drawn with increased detail obtained from visual telescopic observations made by ACIC cartographers at Lowell Observatory.

shadow-measuring program; for this the Aeronautical Chart and Information Center let a contract with the University of Manchester to establish a lunar photographic program at Pic du Midi Observatory. To support this program ACIC provided a USAF K-22 aerial camera which was modified to take lunar exposures. This was mounted on the Pic du Midi 24-in. refractor. The K-22 camera used 9-in. rolls of film which, after exposure, were forwarded to a USAF Base in France, and subsequently flown to ACIC for processing and measurement. In this way ACIC acquired approximately 60 000 time lapsed lunar photographs from 1961 through 1966.

From 1960 to 1967, forty-four LACs were completed. The published charts are indicated on the index shown in Figure 6.5 and listed at the end of this chapter.

3. Ranger Lunar Chart (RLC)

The Ranger Missions were planned to obtain close-up photographs of the lunar surface which would benefit both the scientific program and the manned program. This objective was accomplished by Ranger Missions VII, VIII and IX in 1964–65. These three space-vehicles impacted the Moon in preselected areas to obtain photographs of lunar maria, highlands and highland basins.

From a cartographic viewpoint, the Ranger probes were completely unconventional missions, producing a new type of reconnaissance data. Each Ranger spacecraft contained six television cameras, three of which had lenses of $f/1.0$, 25 mm focal length; the other three, $f/2.0$, 76 mm focal length. The images from these lenses were formed on the photoconductive surface of vidicon tubes, with various combinations of angular coverage, exposure time and scan density.

The exposure and readout of the vidicon images were accomplished by two independent amplification and telemetry chains. One channel consisted of an 'A' camera of 25 mm focal length and a 25° field of view, plus a 'B' camera of 76 mm focal length and 8.4° field. The television signals received from the Ranger transmitter system were displayed on a cathode ray tube and photographed, providing a 35 mm reconstruction of the original televised photograph.

Ranger VII approached the Mare Nubium area from a northwest direction. Impact occurred on July 31, 1964 in the LAC-76 area as indicated in Figure 6.6. The region of impact was subsequently named Mare Cognitum (Sea of Knowledge) by the International Astronomical Union. The velocity vector was within the field of the A camera, and the frames of that camera were nested throughout the mission. The B camera was oriented east of the velocity vector, and its coverage was skewed towards the impact

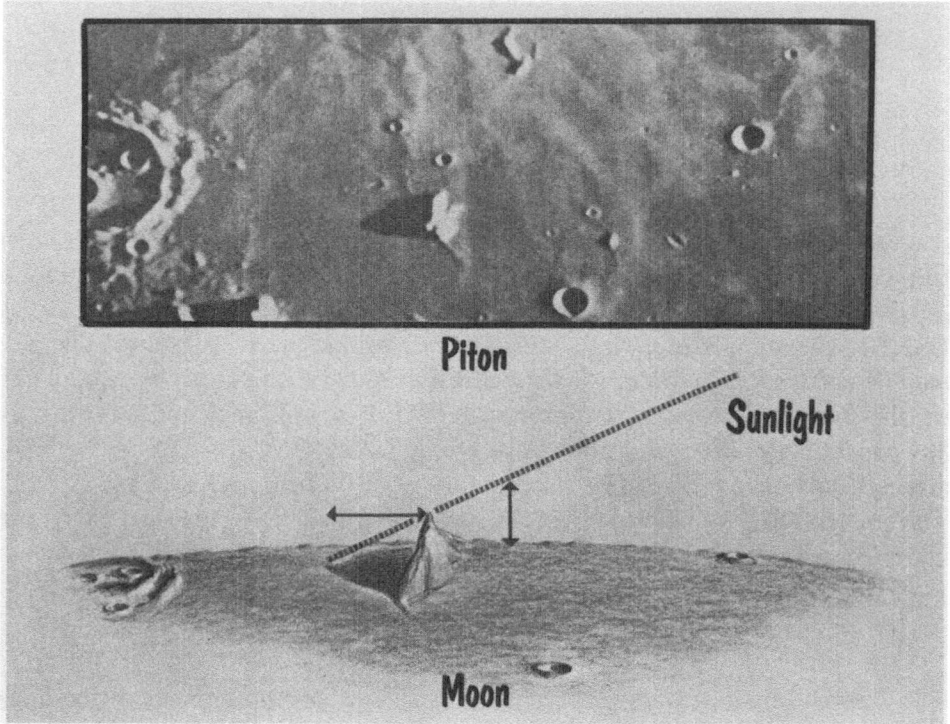

Fig. 6.4. Determining heights of lunar mountains by measuring the length of the shadow cast by the peak dates back to the 18th century. Refined techniques (cf. Chapter 5) developed at Manchester University were used by ACIC to determine relative heights on lunar charts.

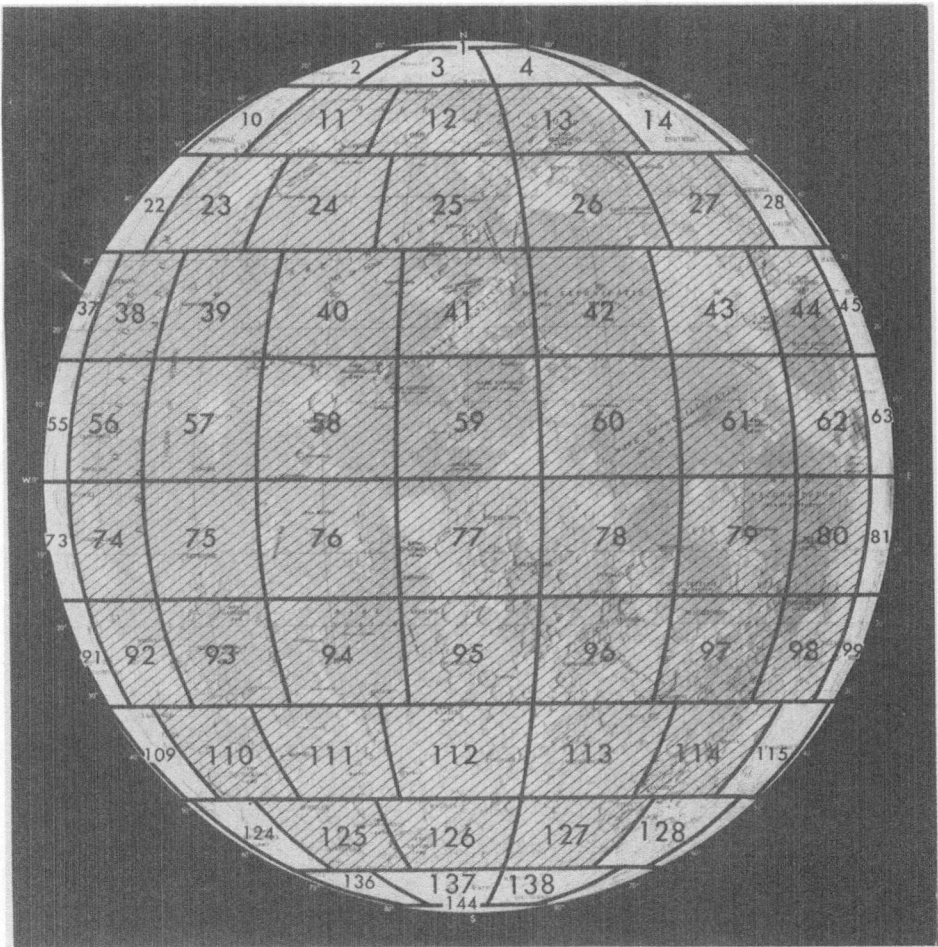

Fig. 6.5.   An index to the 1 : 1 000 000 scale LAC series. Forty-four charts were published by ACIC from 1960 to 1967.

point. Ranger VII transmitted 4038 photographs during the last 17 min and 13 s of its flight.

Ranger VIII impacted the Moon on February 20, 1965 in the southern portion of Mare Tranquillitatis, the region covered by LAC-60. The vehicle approached from the southwest, with all camera axes trailing the velocity vector. As a result, there was no complete nesting of images, but a general skewing towards the impact point. Ranger VIII transmitted 7137 photographs during the last 23 min and 4 s before impact.

Ranger IX struck in the crater Alphonsus located in the LAC-77 area on March 24, 1965. Its trajectory was more nearly vertical than Rangers VII and VIII, with the velocity vector oriented between the A and B camera axes, giving a large amount of nesting and overlap with both sets of images. Ranger IX transmitted 5814 photographs during the last 18 min and 49 s of flight.

After analyzing the Ranger VII film. ACIC decided that the best medium for portrayal of the new information would be a special series of Ranger charts to be identified as the Ranger Lunar Chart (RLC) series. The essential feature of the Ranger coverage

Fig. 6.6.   Robert Bryson, Hq. NASA, is shown pointing to the area on LAC-76 where Ranger VII impacted. Others shown are John Saari, Boeing (left), Robert Carder, ACIC (holding chart), and Richard Shorthill, Boeing (right).   (*Boeing Photo*)

was that of increasing scale and resolution of progressively smaller areas. Therefore, the charts were designed to cover the region in successive increments of scale. It was not felt that any special portrayal was necessary at scales smaller than 1:1 000 000 since Ranger imagery compatible with this type of compilation came from the earliest part of the mission and had poorer resolution than earthbased photographs.

The first chart to reflect new information from Ranger VII was RLC-1, at 1:1 000 000 scale; its chart limits corresponded with LAC-76. Four additional charts, RLC-2, RLC-3, RLC-4 and RLC-5 at scales of 1:500 000; 1:100 000; 1:10 000; and 1:1000 respectively, were compiled with the chart coverage nested towards the impact point similar to the photographs. RLC-3 is shown in Figure 6.7. All five charts of the Ranger VII series were completed and delivered to NASA within three months after impact.

The first stage in the RLC production procedure was the preparation of a set of controlled photomosaics for each chart at the desired scale to use as a compilation base. This work was accomplished under the supervision of ACIC astronomer Harold Spradley. For the 1:1 000 000 scale base (RLC-1), it was necessary to use only the relief drawing from LAC-76 which covered the impact area. For each successive step, the proceeding scale RLC base was enlarged by the appropriate scale factor, and rectified prints of the Ranger photographs were mosaicked to fit the pattern of lunar features, a technique known as a 'bootstrap' operation.

After the mosaic basis were completed, work was begun on the preparation of the cartographic drawings. The first step was to identify lunar features directly on the photomosaics. Experienced scientific illustrators then combined these annotations into shaded relief drawings at each chart scale. When the Ranger charts were finished, they not only contained selenographic drawings, names, relative heights, and scale but also depicted the limits of coverage of the various cameras, ground track of the Ranger spacecraft, and approximate point of impact.

Since each mission had slightly different types of coverage, the chart specifications also varied on a mission-to-mission basis. The Ranger VIII photographs were not nested; therefore, several charts are required for portrayal with optimum continuity of the series. Ranger IX was nested; therefore, these charts are similar to the sequence produced for Ranger VII, making a total of 17 charts covering the RLC series.

It would be misleading to say that these charts contain all the information derived from the Ranger missions. However, they do serve to compress a large part of descriptive data about the impact region into a relatively small number of graphics.

4. Apollo Intermediate Chart (AIC)

In the search for additional planning material for Project Apollo, NASA, in early 1965, required twenty 1:500 000 scale lunar charts covering the equatorial region (see Figure 6.8). The purpose of this series (the Apollo Intermediate Chart) was to facilitate the programming of lunar Orbiter photographs covering potential landing sites near the equator.

The AIC is similar to the LAC. The main difference is in relative flat areas where an attempt was made to increase topographic detail through concentrated visual telescopic observations. AIC-78A (Figure 6.9) covering the Delambre area was published in March 1966. All twenty AIC charts were completed by January 1967.

Relief is expressed by shadient portrayal, relative elevations, crater depths and radius vector values. The relief is printed in duotone black and green, the same as the LAC. On the reverse side of the chart, the relief plate is printed in monochrome black on which selected areas are outlined and described in the margin as interpreted from visual observations.

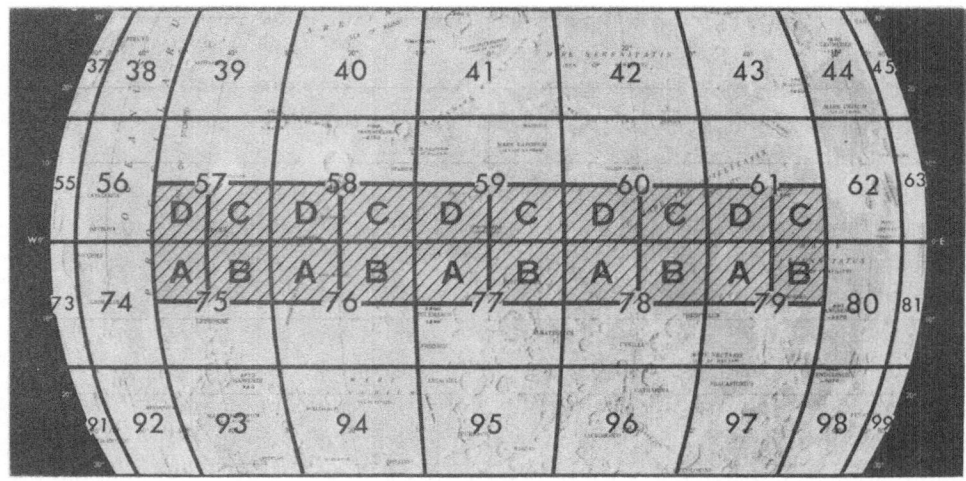

Fig. 6.8.   An index to the 1:500 000 scale Apollo Intermediate Chart (AIC Series). Twenty charts were published by ACIC from 1964 to 1967.

Fig. 6.9.   One of the AIC charts (AIC-78 A) covering the Delambre area and published by ACIC in March 1966.

5. Orbiter Mapping

While the Ranger material gave lunar cartographers a new and closer look at the surface of the Moon, the analysis of this information was still linked with previous techniques and frames of reference. Basic positions were derived from selenocentric control. All topographic analysis was based essentially on monoscopic techniques, since the trajectory did not produce conventional stereoscopic coverage.

The Lunar Orbiter Program, begun by NASA in the summer of 1966, represented a more complete break with the past in terms of photographs obtained, the data reduction techniques, and the cartographic products. This system employed two cameras which used a common roll of film. After the images were exposed, the film was processed in the spacecraft and stored for later readout and transmission.

The Lunar Orbiter would have to be described as a reconnaissance system rather than a mapping tool. Nevertheless, the photographs were more suitable for large-scale topographic mapping than any materials previously used. The two cameras, having 80 mm (medium resolution) and 610 mm (high resolution) focal lengths, were used primarily for vertical photography. The 80-mm camera obtained stereo endlap of as much as 87% between consecutive frames. The 610-mm camera covered a rectangular area in the center of each 80-mm frame, and the maximum resolution equaled or exceeded the best Ranger coverage.

Lunar Orbiter I was launched on August 10, 1966 and placed in an equatorial orbit around the Moon, finally crashing after two and one-half months. During this period, it took a total of 213 medium- and high-resolution photographs. Ten potential Apollo landing areas were photographed and scattered obliques taken of the far side of the Moon and of selected targets on the near side, as a mission 'bonus'. Lunar Orbiters II and III were similarly oriented and provided photographic coverage of other pre-selected areas.

From the returns of the three missions, sixteen lunar landing-site maps were produced in Photomap and Lunar Map form by ACIC and the U.S. Army Map Services (AMS) from March 1967 to June 1968. These sheets are identified by the designator ORB followed by site numbers and an abbreviated statement of scale in parentheses. For example: a 1:100000 scale map of Lunar Orbiter I, Site 5, is designated ORB-I-5 (100). The 1:100000 scale sheets utilized medium-resolution Orbiter photographs, and the 25000 scale sheets were produced from high-resolution photographs. The latter scale was generally produced only in Photomap form.

The Lunar Photomap series reflects only photomosaic detail and is lithographed in black. The Lunar Map series expresses relief by shadient portrayal, contours, spot elevations, height of rim elevations above surrounding terrain and crater depths. Contour values and spot elevations are shown as lunar radius vectors expressed in meters. The contour information was developed by interpolation between analytically developed vertical spot elevations.

The principal deterrent to cartographic exploitation of lunar orbiter photographs was photographic distortion introduced by the scanning and transmission process which segmented each photograph into individual framelets. Generally the displacements are of sufficient magnitude to cause observable discontinuities in images occurring at the edges of adjacent framelets. A pre-exposed film reseau, suggested by ACIC photogrammetrist Duane Lyon, was instituted beginning with the second Orbiter mission; this provided a basis for compensating for distortions introduced by inprocess photo segmentation.

Shown in Figure 6.10 is the first ACIC Orbiter Lunar Map, ORB-I-5 (100) which was published in March 1967 under the supervision of ACIC cartographer Orville Jakubs. Three other Orbiter sites were mapped by ACIC; the other twelve, by AMS.

6. Surveyor Site Mapping

In 1967, NASA requested ACIC to produce large scale maps covering the Surveyor I landing site. Two scales were selected: 1:2000 for a Topomap and Photomap and 1:500 for a Topomap. These maps were compiled from photographs taken from Orbiter missions I and III. The 1:2000 scale Surveyor I site map, published by ACIC in January 1968, is shown in Figure 6.11. The vertical and horizontal control was estab-

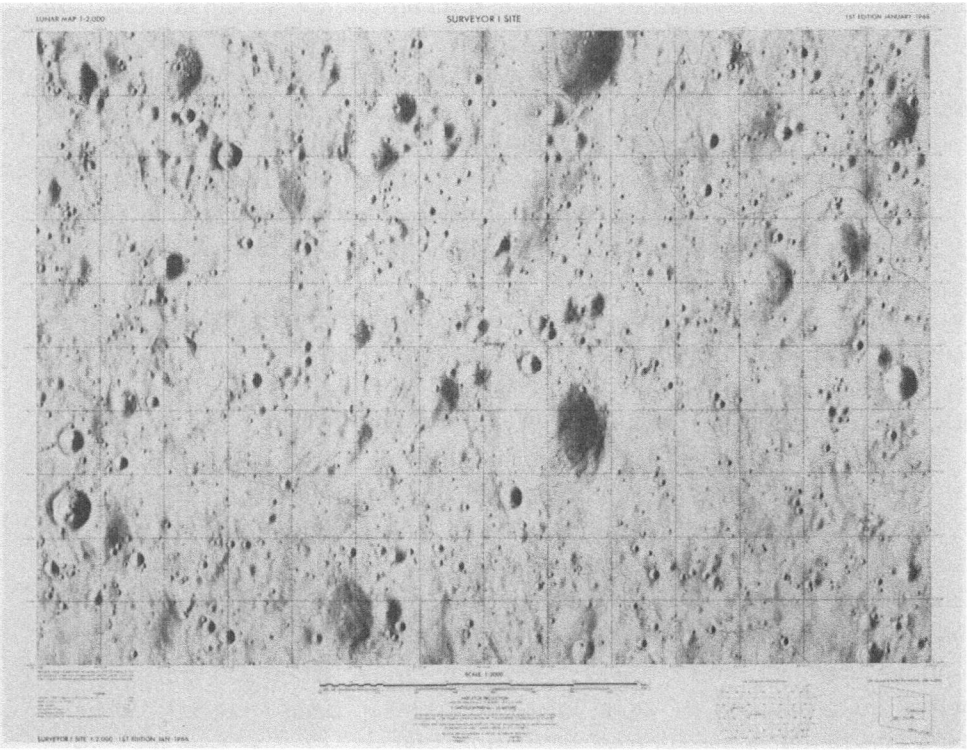

Fig. 6.11. A 1:2000 scale lunar map of the Surveyor I landing area compiled from Orbiter I and III photographs. It was published by ACIC in January 1968.

lished by photogrammetric triangulation using Lunar Orbiter III P-12 control. Contours shown at 10-m intervals were interpolated from vertical control points and printed in red.

7. Scientific Site Mapping

The equatorial orbits of Missions I, II and III were flown over each site at lunar altitude of 46 km, which provided resolutions of 1 to 2 m. Lunar Orbiters IV and V differed in that the spacecrafts were positioned in a near-polar orbit to take advantage of lunar rotation so that successive passes could be spaced to saturate the moon with photographs. From Mission IV the entire near side was photographed at an altitude of 2500 to 3000 km. The final flight, Mission V, was programmed to take areas of scientific interest from an altitude of 100 km.

From Mission V photographs, ACIC mapped two lunar areas of scientific interest, Rima Bode II (Orbiter V, Site 29) and Aristarchus (Orbiter V, Site 48). Three types of maps at 1:250000 scale, Photomap (without contours), Topographic Photomap (with contours), and a Topographic Map were published for Rima Bode II. Only the Topographic Map was produced for Aristarchus. This map, drawn by ACIC cartographer Wayne Kaempfe, is shown in Figure 6.12. It is printed in a duotone green and black with contours at a 400-m interval overprinted in red.

Also ACIC published 1:250000 scale maps of two other areas of scientific interest from Mission III photographs. These were Mösting C (Orbiter III, Site 18) and Fra Mauro (Orbiter III, Site 23). The latter site covered the Apollo 14 landing area.

8. Small Scale Lunar Charting

All Orbiter Missions were programmed to take some photographs of the back side of the Moon. These photographs gave the first real look at far side feature since the U.S.S.R. Zond 3 pictures taken in 1959 were of poor quality. At the conclusion of Orbiter Mission IV, NASA had acquired about 75% coverage of the Moon's far side. With this new material in the spring of 1967, ACIC started work to compile the first extensive chart of the newly discovered far side features.

The Lunar Far Side Chart LFC-1 (1:5000000) and LFC-2 (1:10000000) shown in Figure 6.13, published in August 1967 were presented at the XIII General Assembly of the International Astronomical Union held in Prague, Czechoslovokia. Except for scale the two charts are basically the same. They were developed on a gnomonic projection for the polar areas and a Mercator projection for the central zone. Each projection extends to 48° N S lat., thus allowing feature detail to be matched at this parallel by a rolling fit. Zond 3 photographs were used to extend the equatorial coverage between 105° and 135° W long. However, the low resolution of the Russian photographs caused the rendition in this area to appear sketchy as compared to the Orbiter photographs.

A procedure for establishing the position of features was devised by ACIC photo-

Fig. 6.12.   A 1:250000 scale lunar map of the Aristarchus crater published by ACIC in January 1972
from Orbiter V photographs.

grammetrist Duane Lyon which permitted early completion of LFC-1 and LFC-2. All
photographs were reduced to approximately 1:5000000 and a latitude-longitude
grid was constructed on each image. These perspective projections were extrapolated
from the lunar near side control in the limb regions, together with the center coordinates of the photographs and photo orientation data from the ephemerides. A visual
grid transfer of the photographic detail was made to the chart base, and the features
were pictorially portrayed by shaded relief.

   The position of features on LFC-1 and LFC-2 was considered provisional, having
been independently determined from predicted coordinates of the principal points
of the Lunar Orbiter photographs. Topography was portrayed by airbrush shaded

relief with an assumed lighting from the west. Photographs of the missing areas were acquired by Orbiter V which permitted LFC-1 and LFC-2 to be reissued in October 1967 with complete hemispherical coverage. On the later edition, Orbiter V replaced the previous Zond 3 coverage.

Following the completion of LFC-1 and LFC-2, ACIC began work on compiling a companion near side chart, the *Lunar Earthside Chart*, LEC-1. The 1:5000000 LEC-1 was published in July 1968. The LACs were used as source for this chart and was supplemented by Lunar Orbiter photographs in the limb regions. However, the polar areas for LEC-1 were not compiled because at that time plans were being formulated to completely recompile LEC-1 and LFC-1 based on a new network of control. Thus the ACIC Positional Reference System of 1969 (PRS) devised by cartographer Byron Ruffin and photogrammetrist Lawrence Schimerman of ACIC came

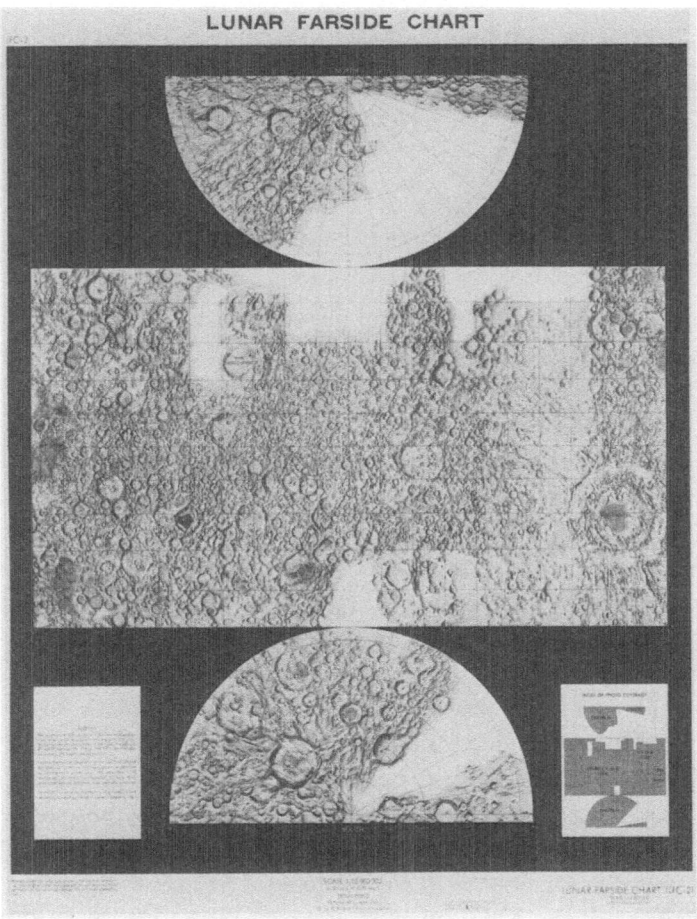

Fig. 6.13.   LFC-2, the first detailed far side chart of the Moon. It was compiled by ACIC in 1967 from Orbiter I, II, III, IV and Zond 3 photographs. The areas shown in white were completed at a later date from Orbiter V photographs.

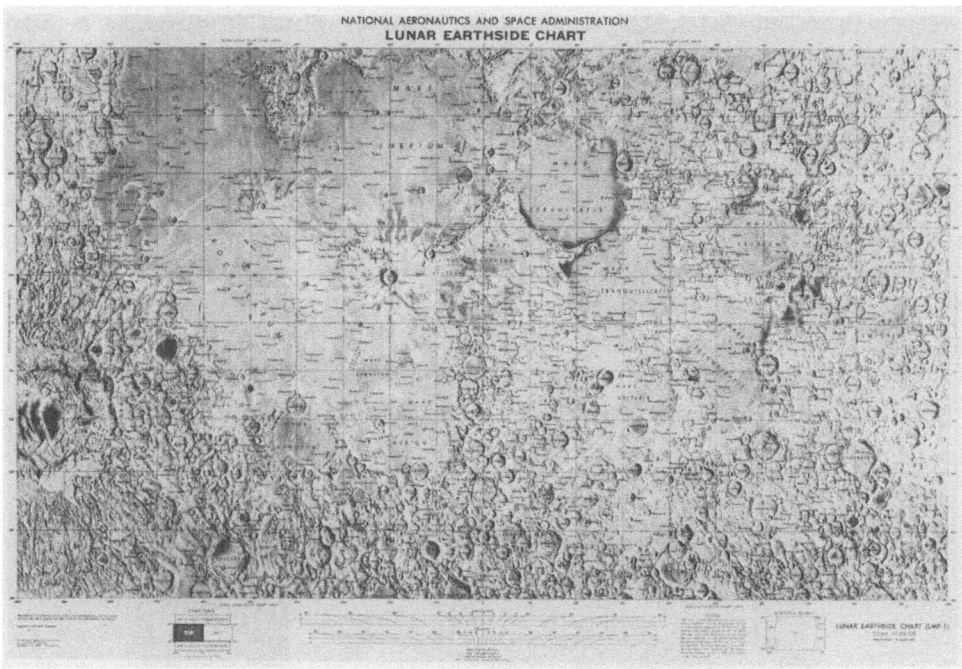

Fig. 6.14.   Lunar Earthside Chart (LMP-1).

into being and formed the basic control network for a new 1:5000000 scale series named *Lunar Earthside Chart*, LMP-1 (Figure 6.14), *Lunar Farside Chart* LMP-2 (Figure 6.15) and *Lunar Polar Chart.* LMP-3 (Figure 6.16).

The PRS consisted of using orbital data to compute perspective projections of selected Lunar Orbiter photographs where overlap coverage was available. The projections were linked into an equatorial band and a meridional arc to provide extensions of coordinates from the lunar near side to the lunar far side (Figure 4.3).

The LMP charts provide complete coverage of the lunar sphere and serve as a basic reference planning series. In the compilation process, Orbiter photographs were reduced, rectified and paneled to the PRS control. This formed the base for drawing the shaded relief. In the shaded relief rendering, the conventional west lighting was changed to an east lighting to closely approximate shadows as would be seen on the Earth side of the Moon during an Apollo mission flight. The relief was printed in brown, highlighted by ray patterns and albedo background printed in blue.

The LMP series was first published in January 1970, but without names for the far side features. Following the XIV IAU General Assembly in August 1970, a second edition was issued in October 1970 which included 513 newly approved names.

The availability of the LMP series in January 1970 afforded an opportunity to publish a new 1:10000000 scale lunar chart. Therefore, LPC-1 produced from LMP-1, -2 and -3, shown in Figure 6.17, came into being in March 1970. In assembling LPC-1, it was decided to center the equatorial area on the first meridian. This decision

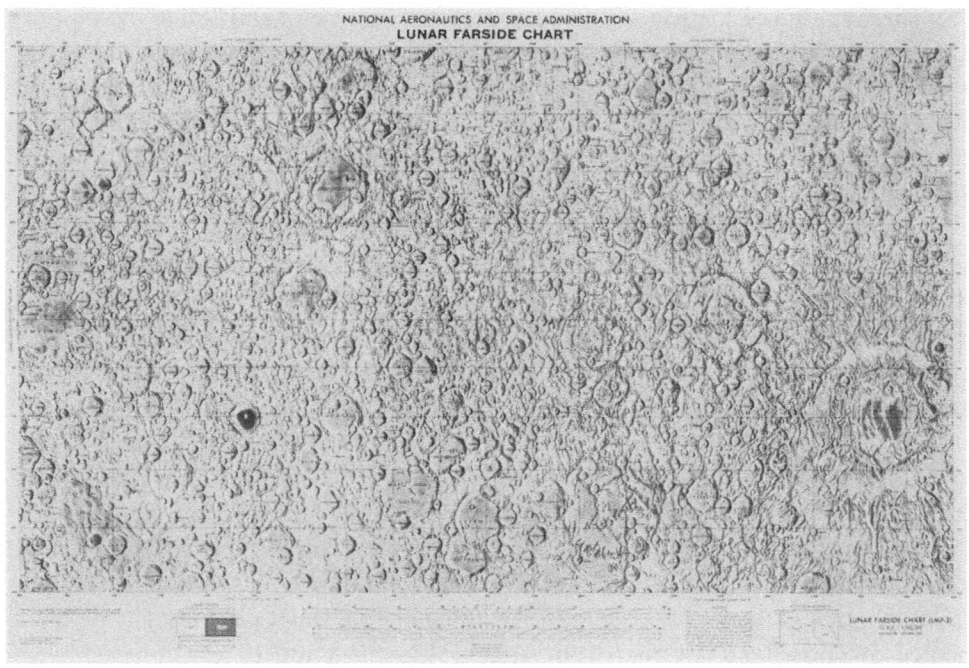

Fig. 6.15.   Lunar Farside Chart (LMP-2).

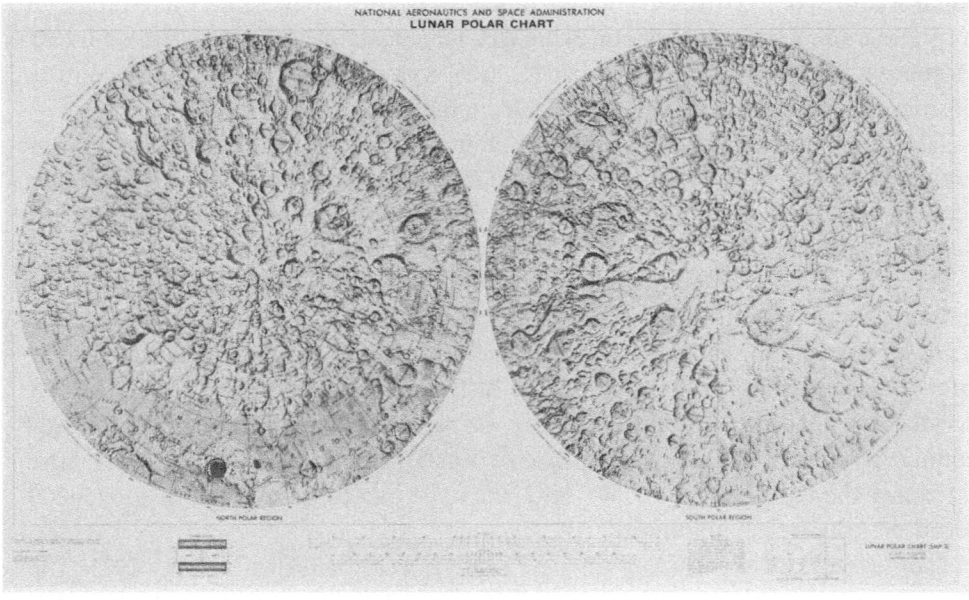

Fig. 6.16.   Lunar Polar Chart (LMP-3).

resulted in splitting LMP-2 at 180° and adding half to the east and half to the west of LMP-1 to form LPC-1. Printed in brown and blue against a black background LPC-1 is similar to LFC-2. All names of primary features were included with the exception of the far side names which had not been approved at that time. In October 1970, when the new far side names became available, NASA decided not to reissue LPC-1 with the new names.

A special NASA edition of LPC-1 was issued in August 1970 in support of the XIV IAU General Assembly held in Brighton, England. Unofficial new names proposed by the IAU Nomenclature Committee were overprinted in purple. This special edition was printed in limited quantities and only file copies now remain.

By far the most comprehensive mapping from Orbiter photographs was the compilation of a special series of four 1:2 500 000 scale equatorial charts identified as the LOC Planning Series (LOC-1, -2, -3 and -4). The LOCs were designed to (1) cover the Apollo area of interest; (2) provide a greater density of feature detail than shown on the 1:5 000 000 LMP series; (3) show all primary and secondary approved IAU names; and (4) serve as a common base for the various Apollo Operational Mission Charts.

Compilation of the LOCs began in 1968 based upon Orbiter photographs supplemented by Hasselblad coverage from Apollo Missions 8 and 10. Earth-based telescopic full-moon photographs were used as source for albedo patterns and ray structure. Details from the Lunar Astronautical Charts (LACs) were also used to the extent of their coverage.

The ACIC Provisional Reference System of 1969 was used for control, supplemented by a NASA Manned Spacecraft Center (MSC) local-control strip in the equatorial area of 72° W to 56° E long. This control, prepared by MSC, was established by photogrammetric triangulation using Orbiter constraints and was based on Lunar Orbiter IV ephemeris data of October 15, 1968. As a result of using the MSC control, positions in this area on the LOCs differ from the LMPs some 3 to 5 km.

The compilation procedure required the mosaicking of rectified Orbiter photography to a control plot. This resulted in a controlled photo base from which lunar relief was rendered in third dimension by airbrush techniques using an assumed east lighting. The final charts were lithographed in brown, blue and black.

The original LOCs, published in July 1969, were limited to 25° N–S lat. and only three charts were completed (LOCs -2, -3 and -4). Numbers were used to designate lunar far side features since IAU names for this area were not approved until a year later. Before starting production on the last chart in this series (LOC-1), NASA decided to extend the latitude coverage to 40° N–S. Work was then directed towards adding another 15° N–S to the existing charts and producing LOC-1. However, after extending the latitude limits to 40° N–S, the scale had to be reduced to 1:2 750 000 because of press-size limitations.

The revised LOCs, including the new LOC-1, were published in May 1971. The litho colors are the same as those on the original LOCs; however, the shaded relief was enhanced by printing in duotone brown. Shown in Figure 6.18 is LOC-2, which

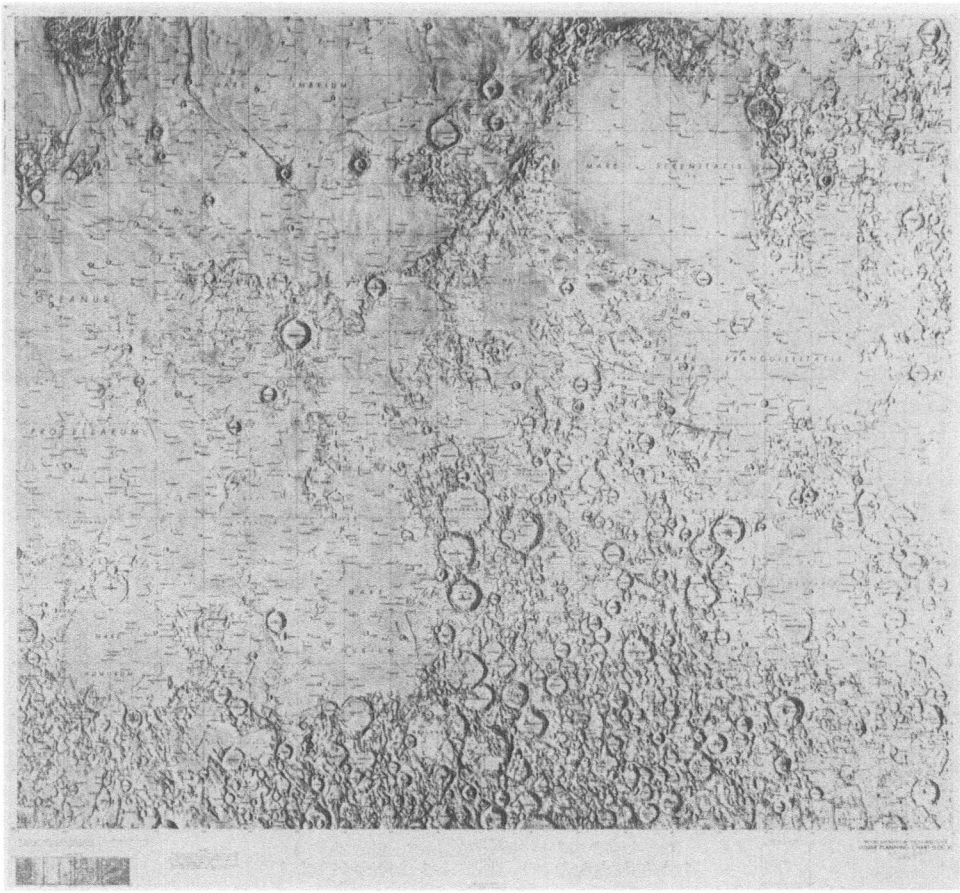

Fig. 6.18.   One of four Lunar Planning Charts (LOC-2), scale 1:2 750 000 covering the equatorial region
from 40° N to 40° S latitude. LOC-2 was published by ACIC in May 1971.

was published in May 1971. ACIC personnel primarily responsible for the LOC series
were Jerry Higgins, Charles Moore, Raymond Anderson, Wayne Kaempfe and Fred
Paris.

9. Apollo Site Mapping

Apollo Missions 15, 16 and 17 provided high-quality mapping photographs along
with precise supporting data for about 20% of the lunar surface. The instruments
flown comprised a metric camera system and a panoramic camera.

The metric camera system consists of a mapping camera with a 76 mm $f/4.5$ lens,
associated stellar camera, laser altimeter and precise timing equipment. Inscribed on
the focal plane platen of the mapping camera is a 10-mm reseau which is recorded on
each photograph by natural illumination; also inscribed are artifically illuminated
fiducial marks. This camera is designed to compensate for forward motion during the

exposure sequence and to automatically set the duration of exposure; it obtained a ground resolution of 20 m. To provide stereoscopic coverage it was flown with a 78% overlap between succeeding frames. These photographs are best suited for stereo-compiling topographic maps at scales of 1:250000 and smaller.

The panoramic camera, using a 610 mm $f/3.5$ lens obtained high-quality 2- to 3-m ground-resolution stereoscopic coverage of a broad swath along the entire daylight portion of a number of orbital tracks on each mission. Because of the high-quality resolution, these photographs are best suited for compiling maps at 1:50000 scale and larger. Shown in Figure 6.19 is a small section of a panoramic photograph taken of the Apollo 15 landing area. This image, greatly magnified, shows the Lunar Module (LM) resting on the lunar surface. Easily identified is the LM shadow and three of its footpads.

ACIC was the first to compile a precision lunar map from these high-quality mapping photographs. The area, the Apollo 15 landing site, was produced at 1:25000 on a transverse Mercator projection with 20-m contours supplemented with a 10-m interval in the flat plain.

The compilation was produced on an AS-11B1 analytical stereoplotter which has

Fig. 6.19.   Arrow points to the Apollo 15 Lunar Module resting on the lunar surface at Hadley Rille. This photograph was taken by the Apollo 15 610-mm panoramic camera from an altitude of 60 nm.

unique properties that enable it to handle many non-standard photogrammetric applications. This plotter is capable of handling a wide range of photographs from vertical frame to convergent panoramic. The convergent panoramic photographs do not need to be transformed or rectified before they are used in the AS-11B1 plotter. This plotter can also correct for lunar curvature, film shrinkage, lens distortion, image motion, vehicle motion and stereo model deformation due to distortions.

Figure 6.20 shows a section of the published 1:25000 scale Lunar Topographic

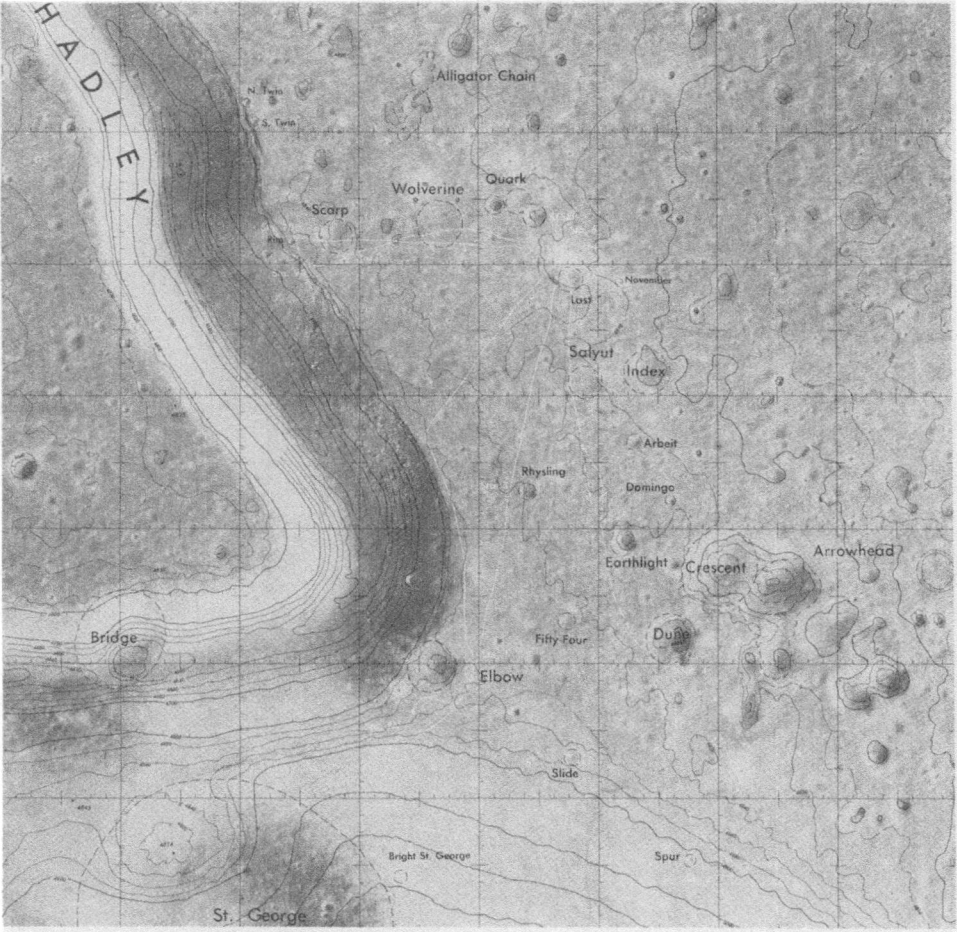

Fig. 6.20.   This is a section of the 1:25000 scale Lunar Topographic Photomap covering the Apollo 15 landing site. The contours were compiled on an AS 11B1 analytical steroplotter by ACIC in February 1972.

Photomap, Rima Hadley, issued in February 1972. The photo image is printed in halftone black with the contours overprinted in red. The crater names were temporarily assigned by NASA to facilitate communications during the mission and are not approved by the IAU.

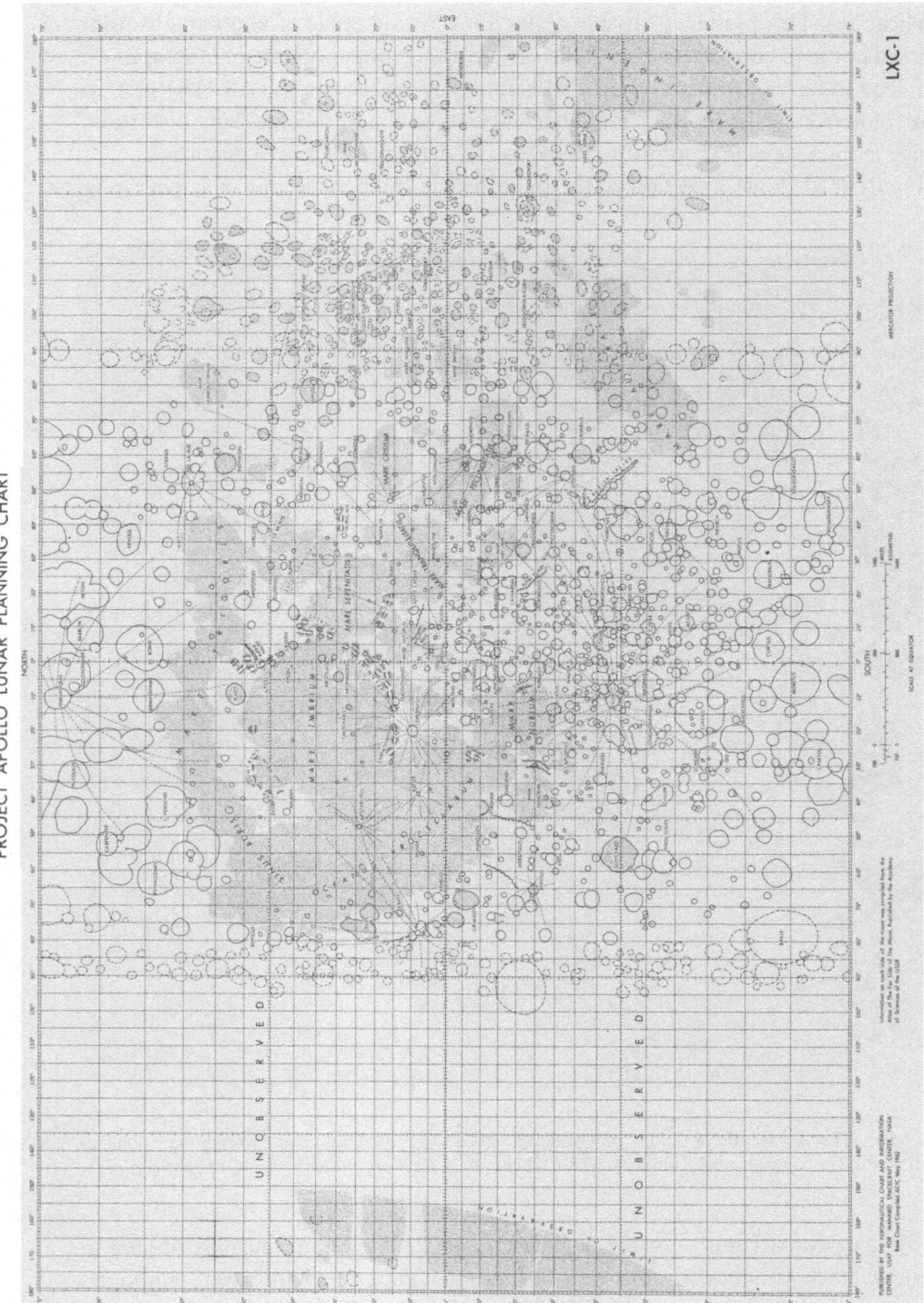

Fig. 6.21.   The Project Apollo Lunar Planning Chart produced on a Mercator projection by ACIC in May 1962.

10. Special Lunar Charts

In May 1962, ACIC produced for NASA a special 1:21 000 000 scale plotting chart of the Moon on a Mercator projection from 75° N to 75° S lat. To the best of our knowledge this was the first chart of the Moon compiled on the Mercator projection. The Mercator projection served to examine the relationship of orbit angle inclination to the lunar surface and to plan lunar orbits prior to visual reconnaissance and manned landings.

This chart titled *Project Apollo Lunar Planning Chart*, LXC-1, 24″ × 16″, is shown in Figure 6.21. LXC-1 is a schematic drawing which shows major craters by outline, rays by dashed lines, mountain ranges by hachures and the maria by a dot stipple. Information for the far side was compiled from the U.S.S.R. Far Side Map produced from Luna 3 photographs taken in 1959. (Complete coverage of the far side was not available at that time.)

Also in 1966–67, ACIC produced several other special lunar charts worthy of note.

Fig. 6.22.   Lunar Isothermal Chart (LIC-1).

One was the Lunar Isothermal Chart (LIC-1) shown in Figure 6.22. This chart consists of the Lunar Reference Mosaic LEM-1, printed in gray and overprinted with red contours representing temperature levels in deg. Kelvin. Source for the thermal data was furnished by Richard Shorthill and John Saari of the Boeing Scientific Research Laboratories, Seattle, Wash. (see Figure 6.6). They acquired the data by rapidly scanning the Moon with an infrared sensor attached to the 60-in. Mt. Wilson telescope.

When the infrared observations were made September 26, 1963, the mean subsolar point was 1°29′24″ N, 82°47′24″ E, which limited the observational range to the eastern half of the lunar Earth-side hemisphere. In other words, the Moon's phase on that date was about 8 days old with the terminator falling at approximately 8° west longitude.

The thermal contours are expressed in units of two from 1 through 100 with 1 representing a low temperature of 137.5 K and 100 a high temperature of 366.8 K. To assist in interpretation, five different tint bands, each representing approximately 50 K, were overprinted in colors from yellow through orange.

Fig. 6.23.   Lunar Albedo Reference Chart (ARC-1).

While information is available to produce a series of Lunar Isothermal Charts at different phase angles, LIC-1 is the only chart of this type published.

Another special chart is the Lunar Albedo Reference Chart, ARC-1 shown in Figure 6.23. ARC-1, compiled under the direction of ACIC astronomer Harold Spradley consists of the Lunar Reference Chart LEM-1, printed in gray and over-

Fig. 6.24.   The 16-in. diam. NASA Lunar Globe produced by ACIC in May 1969.

Fig. 6.25.   Scientific illustrator Jay Inge using the airbrush to draw 30° gores for the NASA Lunar Globe.

printed with red contours representing levels of equal brightness, as derived from astronomical observations made near full Moon. While it is impossible to measure the true 'normal albedo' of the lunar surface from the Earth, these observations are valuable in describing radiance differences which are primarily albedo-dependent.

Photometric calibration was provided through the exposure of a sensitometric step wedge along with the lunar image prior to photographic processing. From the derived characteristic curves, relative exposure values were recaptured with a deviation of less than one percent of the average exposure level. Each photograph was measured and densities were converted to exposure values through individual wedges. Thus, the contours represent interpolated levels of relative brightness when the lunar disk is illuminated and viewed normal at the unlibrated 0° phase. The contours of equal brightness shown on ARC-1 are expressed in 1° differences.

Ten Pic du Midi Observatory full-moon photographs were used as source. One photograph was selected as a base and measurements from the other nine were then normalized to the reference photograph.

11. NASA Lunar Globe

In May 1969, ACIC published the 16-in. diam NASA Lunar Globe shown in Figure

6.24. This 1:8 533 150 scale globe and mounting was designed by ACIC cartographer Howard Holmes. The stand has a metal base with a simulated wood grain finish and a movable calibrated equatorial ring which permits coordinate determinations and great-circle measurements.

The 30° gore drawings for the NASA globe were produced at the ACIC Lowell office by Lowell Observatory scientific illustrator Jay Inge (see Figure 6.25). The far side features were drawn from the Orbiter photographs and the near side detail came from LAC drawings plus earth-based telescopic photographs. Surface features were drawn in color with an assumed eastern light source to approximate prevailing conditions during a morning descent and landing by spacecrafts.

The projection interval is 10° in latitude and 15° in longitude. All primary IAU official names are shown on the near side but only four of the far side names could be positively identified at the time this globe was compiled.

The printing of the gores in gray-green and the manufacture of the globe was accomplished under a commercial contract with the Denoyer-Geppert Company, Chicago, Illinois.

References

Carder, Robert W.: 1961, 'Photo Topography for the 1:1 000 000 Lunar Charts', Photogrammetric Engineering: *Journal of the American Society of Photogrammetry*, June.

Carder, Robert W.: 1962, 'Air Force Maps of the Moon', *The Review of Popular Astronomy*, May–June.

Jakubs, Orville S.: 1966, 'Transitions in Lunar Cartography', Paper presented at the Annual ASP/ACSM Convention, March 6–11.

Spradley, Harold L.: 1965, 'Lunar Charting from Ranger Photography', Paper presented at the Semi-Annual ASP Convention, September 22–24.

LUNAR MAPS PRODUCED BY U.S. AIR FORCE FOR NASA

Description	Sheet No.	Sheet name	Scale	Total sheets	Date	Sheet size
Lunar Mosaics						
	LEM-1A	USAF Lunar Reference Mosaic	1:10000000	1	2-60	17″ × 18″
	LEM-1	USAF Lunar Reference Mosaic	1:5000000	1	2-60	34″ × 35″
	LEM-1B	USAF Lunar Wall Mosaic	1:2500000	2	1-63	58″ × 70″
Lunar Astronautical Charts (LAC Series)						
	LAC-11	J. Herschel	1:1000000	1	3-67	22″ × 29″
	LAC-12	Plato	1:1000000	1	1-67	22″ × 29″
	LAC-13	Aristoteles	1:1000000	1	7-67	22″ × 29″
	LAC-23	Rümker	1:1000000	1	2-67	22″ × 29″
	LAC-24	Sinus Iridum	1:1000000	1	9-66	22″ × 29″
	LAC-25	Cassini	1:1000000	1	9-66	22″ × 29″
	LAC-26	Eudoxus	1:1000000	1	3-67	22″ × 29″
	LAC-27	Geminus	1:1000000	1	7-67	22″ × 29″
	LAC-38	Seleucus	1:1000000	1	3-65	22″ × 29″
	LAC-39	Aristarchus	1:1000000	1	11-63	22″ × 29″
	LAC-40	Timocharis	1:1000000	1	10-63	22″ × 29″
	LAC-41	Montes Apenninus	1:1000000	1	9-63	22″ × 29″
	LAC-42	Mare Serenitatis	1:1000000	1	2-65	22″ × 29″
	LAC-43	Macrobius	1:1000000	1	5-65	22″ × 29″
	LAC-44	Cleomedes	1:1000000	1	12-65	22″ × 29″
	LAC-56	Hevelius	1:1000000	1	5-63	22″ × 29″
	LAC-57	Kepler	1:1000000	1	5-62	22″ × 29″
	LAC-58	Copernicus	1:1000000	1	7-61	22″ × 29″
	LAC-59	Mare Vaporum	1:1000000	1	4-63	22″ × 29″
	LAC-60	Julius Caesar	1:1000000	1	9-62	22″ × 29″
	LAC-61	Taruntius	1:1000000	1	2-63	22″ × 29″
	LAC-62	Mare Undarum	1:1000000	1	2-64	22″ × 29″
	LAC-74	Grimaldi	1:1000000	1	4-62	22″ × 29″
	LAC-75	Letronne	1:1000000	1	6-62	22″ × 29″
	LAC-76	Montes Riphaeus	1:1000000	1	4-64	22″ × 29″
	LAC-77	Ptolemaeus	1:1000000	1	5-63	22″ × 29″
	LAC-78	Theophilus	1:1000000	1	3-63	22″ × 29″
	LAC-79	Colombo	1:1000000	1	4-63	22″ × 29″

LUNAR MAPS PRODUCED BY U.S. AIR FORCE FOR NASA *(continued)*

Description	Sheet No.	Sheet name	Scale	Total sheets	Date	Sheet size
Lunar Astronautical Charts (LAC Series)						
	LAC-80	Langrenus	1:1 000 000	1	3-64	22″ × 29″
	LAC-92	Byrgius	1:1 000 000	1	2-66	22″ × 29″
	LAC-93	Mare Humorum	1:1 000 000	1	6-62	22″ × 29″
	LAC-94	Pitatus	1:1 000 000	1	5-64	22″ × 29″
	LAC-95	Purbach	1:1 000 000	1	12-64	22″ × 29″
	LAC-96	Rupes Altai	1:1 000 000	1	4-65	22″ × 29″
	LAC-97	Fracastorius	1:1 000 000	1	5-65	22″ × 29″
	LAC-98	Petavius	1:1 000 000	1	5-66	22″ × 29″
	LAC-110	Schickard	1:1 000 000	1	9-67	22″ × 29″
	LAC-111	Wilhelm	1:1 000 000	1	10-67	22″ × 29″
	LAC-112	Tycho	1:1 000 000	1	7-67	22″ × 29″
	LAC-113	Maurolycus	1:1 000 000	1	12-66	22″ × 29″
	LAC-114	Rheita	1:1 000 000	1	10-66	22″ × 29″
	LAC-125	Schiller	1:1 000 000	1	10-67	22″ × 29″
	LAC-126	Clavius	1:1 000 000	1	10-67	22″ × 29″
	LAC-127	Hommel	1:1 000 000	1	11-67	22″ × 29″
Ranger VII Charts						
	RLC-1	Mare Cognitum	1:1 000 000	1	10-64	22″ × 29″
	RLC-2	Guericke	1:500 000	1	10-64	22″ × 29″
	RLC-3	Bonpland H	1:100 000	1	10-64	22″ × 29″
	RLC-4	Bonpland PQC	1:10 000	1	10-64	22″ × 29″
	RLC-5	Unnamed	1:1000	1	10-64	22″ × 29″
Ranger VIII Charts						
	RLC-6	Hypatia	1:1 000 000	1	3-66	22″ × 29″
	RLC-7	Sabine	1:250 000	1	3-66	22″ × 29″
	RLC-8	Sabine D	1:100 000	1	3-66	22″ × 29″
	RLC-9	Sabine DM	1:50 000	1	3-66	22″ × 29″
	RLC-10	Sabine EF	1:15 000	1	3-66	22″ × 29″
	RLC-11	Sabine EB	1:5000	1	3-66	22″ × 29″
	RLC-12	Sabine EBF	1:2000	1	3-66	22″ × 29″

LUNAR MAPS PRODUCED BY U.S. AIR FORCE FOR NASA (*continued*)

Description	Sheet No.	Sheet name	Total sheets	Scale	Date	Sheet size
Ranger IX Charts						
	RLC-13	Ptolemaeus	1	1:1 000 000	5-66	22″ × 29″
	RLC-14	Alphonsus	1	1:250 000	5-66	22″ × 29″
	RLC-15	Alphonsus GA	1	1:50 000	5-66	22″ × 29″
	RLC-16	Alphonsus GP	1	1:10 000	5-66	22″ × 29″
	RLC-17	Alphonsus GLH	1	1:2000	5-66	22″ × 29″
Apollo Intermediate Charts (AIC Series)						
	AIC-57C	Encke	1	1:500 000	8-66	22″ × 29″
	AIC-57D	Maestlin	1	1:500 000	8-66	22″ × 29″
	AIC-58C	Gambart	1	1:500 000	8-66	22″ × 29″
	AIC-58D	Reinhold	1	1:500 000	8-66	22″ × 29″
	AIC-59C	Triesnecker	1	1:500 000	1-66	22″ × 29″
	AIC-59D	Pallas	1	1:500 000	1-66	22″ × 29″
	AIC-60C	Arago	1	1:500 000	3-66	22″ × 29″
	AIC-60D	Agrippa	1	1:500 000	8-66	22″ × 29″
	AIC-61C	Secchi	1	1:500 000	1-67	22″ × 29″
	AIC-61D	Maskelyne D	1	1:500 000	5-66	22″ × 29″
	AIC-75A	Flamsteed	1	1:500 000	8-66	22″ × 29″
	AIC-75B	Wichmann	1	1:500 000	8-66	22″ × 29″
	AIC-76A	Euclides P	1	1:500 000	6-66	22″ × 29″
	AIC-76B	Fra Mauro	1	1:500 000	6-66	22″ × 29″
	AIC-77A	Flammarion	1	1:500 000	9-66	22″ × 29″
	AIC-77B	Hipparchus	1	1:500 000	3-66	22″ × 29″
	AIC-78A	Delambre	1	1:500 000	3-66	22″ × 29″
	AIC-78B	Torricelli	1	1:500 000	4-66	22″ × 29″
	AIC-79A	Capella	1	1:500 000	6-66	22″ × 29″
	AIC-79B	Messier	1	1:500 000	12-66	22″ × 29″
Orbiter Site Mapping						
Lunar Photomap	ORB-I-5 (100)	—	1	1:100 000	3-67	19″ × 40″
	ORB-I-9.2 (25)	—	8	1:25 000	3-67	23″ × 30″
	ORB-I-9.2 (100)	—	1	1:100 000	3-67	24″ × 40″

LUNAR MAPS PRODUCED BY U.S. AIR FORCE FOR NASA (continued)

Description	Sheet No.	Sheet name	Total sheets	Scale	Date	Sheet size
	ORB-II-8 (25)	—	4	1:25000	10-67	30" × 36"
	ORB-II-8 (100)	—	1	1:100000	8-67	27" × 27"
	ORB-III-12 (25)	—	9	1:25000	6-68	29" × 42"
	ORB-III-12 (100)	—	1	1:100000	1-68	26" × 47"
Orbiter Site Mapping						
Lunar Map	ORB-1-5 (100)		1	1:100000	3-67	19" × 40"
	ORB-1-9.2 (25)		8	1:25000	5-67	23" × 30"
	ORB-1-9.2 (100)		1	1:100000	4-67	23" × 30"
	ORB-II-8 (25)		4	1:25000	2-68	31" × 36"
	ORB-II-8 (100)		1	1:100000	12-67	26" × 28"
	ORB-III-12 (100)		1	1:100000	3-68	36" × 47"
Surveyor Site Mapping						
Surveyor I Site – Lunar Map		—	1	1:2000	1-68	24" × 30"
Surveyor I Site – Lunar Photomap		—	1	1:2000	1-68	24" × 30"
Surveyor I Site – Lunar Map		—	1	1:500	1-68	24" × 30"
Scientific Site Mapping						
Lunar Photomap	Orbiter III-Site 18	Mösting C	2	1:25000	12-69	28" × 44"
	Orbiter III-Site 18	Mösting C	1	1:250000	8-69	19" × 19"
	Orbiter III-Site 23	Fra Mauro	1	1:25000	8-69	24" × 43"
	Orbiter III-Site 23	Fra Mauro	1	1:250000	6-69	19" × 19"
	Orbiter V-Site 29	Rima Bode II	1	1:250000	11-69	18" × 22"
Lunar Topographic Photomap	Orbiter III-Site 18	Mösting C	1	1:250000	12-69	19" × 19"
	Orbiter III-Site 23	Fra Mauro	1	1:250000	12-69	19" × 19"
	Orbiter V-Site 29	Rima Bode II	1	1:250000	12-69	18" × 22"
Lunar Topographic Map	Orbiter III-Site 23	Fra Mauro	1	1:250000	4-70	19" × 19"
	Orbiter V-Site 29	Rima Bode II	1	1:250000	8-70	18" × 22"
	Orbiter V-Site 48	Aristarchus	1	1:250000	1-72	21" × 32"
Small Scale Lunar Charts						
	LFC-1 (1st Ed)	Lunar Farside Chart	1	1:5000000	8-67	27" × 39"
	LFC-1 (2nd Ed)	Lunar Farside Chart	1	1:5000000	10-67	27" × 39"

LUNAR MAPS PRODUCED BY U.S. AIR FORCE FOR NASA (continued)

Description	Sheet No.	Sheet name	Scale	Total sheets	Date	Sheet size
	LFC-2	Lunar Farside Chart	1:10000000	1	8-67	23" × 29"
	LEC 1	Lunar Earthside Chart	1:5000000	1	7-68	27" × 42"
	LMP-1 (1st Ed)	Lunar Earthside Chart	1:5000000	1	1-70	29" × 41"
Small Scale Lunar Charts						
	LMP-1 (2nd Ed)	Lunar Earthside Chart	1:5000000	1	10-70	29" × 41"
	LMP-2 (1st Ed)	Lunar Farside Chart	1:5000000	1	1-70	29" × 41"
	LMP-2 (2nd Ed)	Lunar Farside Chart	1:5000000	1	10-70	29" × 41"
	LMP-3 (1st Ed)	Lunar Polar Chart	1:5000000	1	1-70	29" × 47"
	LMP-3 (2nd Ed)	Lunar Polar Chart	1:5000000	1	10-70	29" × 47"
	LPC-1	Lunar Chart	1:10000000	1	3-70	26" × 38"
	LOC-1	Lunar Planning Chart	1:2750000	1	5-71	41" × 44"
	LOC-2	Lunar Planning Chart	1:2750000	1	5-71	41" × 44"
	LOC-3	Lunar Planning Chart	1:2750000	1	5-71	41" × 44"
	LOC-4	Lunar Planning Chart	1:2750000	1	5-71	41" × 44"
Apollo Site Mapping Lunar Topographic Photomap		Rima Hadley	1:25000	1	2-72	43" × 59"
Special Lunar Charts						
	LIC-1	Lunar Isothermal Chart	1:5000000	1	1-66	34" × 35"
	ARC-1	Lunar Albedo Reference Chart	1:5000000	1	3-67	34" × 35"
	LXC-1	Apollo Lunar Planning Chart	1:21000000	1	5-62	24" × 16"

CHAPTER 7

LUNAR MAPPING AT LOWELL OBSERVATORY

From September 1961 through April 1969 the USAF Aeronautical Chart and Information Center (ACIC) maintained a staff of cartographic visual observers, scientific illustrators and photographers at Lowell Observatory to compile detailed lunar charts from visual telescopic observations.

Lowell Observatory is situated on a mesa one mile west and about 300 ft above the city of Flagstaff. Covering about 700 acres in the midst of the great pine forest of Northern Arizona, the observatory is in the shelter of the 12000-ft San Francisco peaks, just north of Flagstaff. The location is exceptionally favorable for astronomical work because of the large number of cloudless nights each year and the steadiness of its atmosphere. Its altitude, 7250 ft above sea level, places it well above the dust and haze of the lower elevations.

The 24-in. Clark refractor is housed in its original wooden dome built in 1896 and shown in Figure 7.1. The story is told that when Percival Lowell had it built he floated the dome on wooden pontoons in a water trough lined with copper. Salt was added to the water to keep it from freezing but strong winds sprayed salt water on the instruments; the salt also reacted with the copper. Lowell soon abandoned this idea and mounted the dome on railroad-car wheels, later to be replaced with automobile rubber-tire wheels which are still in use.

The 24-in. telescope shown in Figure 7.2 is considered to have excellent optics with a $f/16$ objective lens. The lens is corrected for the visual region of the spectrum, which is roughly between 5000 and 6200 Å. The visual resolution of this lens is approximately 1/10 of a second of arc which means that one can visually observe a lunar crater less than 600 ft in diam, provided atmospheric conditions are steady and transparent.

1. Visual vs Photographic Observations

In 1959, when ACIC embarked on a program to map the Moon, the very best lunar photographs were used as source for the relief portrayal. The first ACIC lunar drawings were reviewed by Gerard Kuiper, then the director of Yerkes Observatory and a lifelong visual observer of the Moon. He advised that it was impossible to interpret fine structure of lunar relief from photographs without visual observations because one can see visually two to three times more than can be seen on the best photograph.

The truth of this statement was dramatically revealed when Kuiper invited ACIC to visually observe the Moon with the 40-in. Yerkes refractor. The need for a telescope was quite apparent when visual observations gave the drawings more detailed and definitive interpretations than could be obtained from photographs alone. This is illustrated in the drawing of Gassendi, shown in Figure 7.3, where detail such as rilles, small craters and hummocks could not be photographed through the telescope

Fig. 7.1.   This unique wooden structure, built in 1896, houses the Lowell 24-in. refractor. The shutter doors in the dome are opened and closed by a series of ropes and pulleys.   *USAF Photo*

but were clearly seen by visual observers. The reason is that atmospheric turbulence called 'seeing' causes the image to be in motion, distorted and defocused.

E. C. Slipher, the noted Mars observer once remarked, "If you expose a photographic plate for two seconds to get a certain image, the photograph turns out to be a bad average over the whole period of time; whereas, the eye can see in 1/20 of a

Fig. 7.2. The Lowell Observatory 24-in. refractor. Its $f/16$ lens, manufactured by Clark, is still one of the finest in the world. Three other refracting telescopes are fixed on the 24-in. tube. *Lowell Photo*

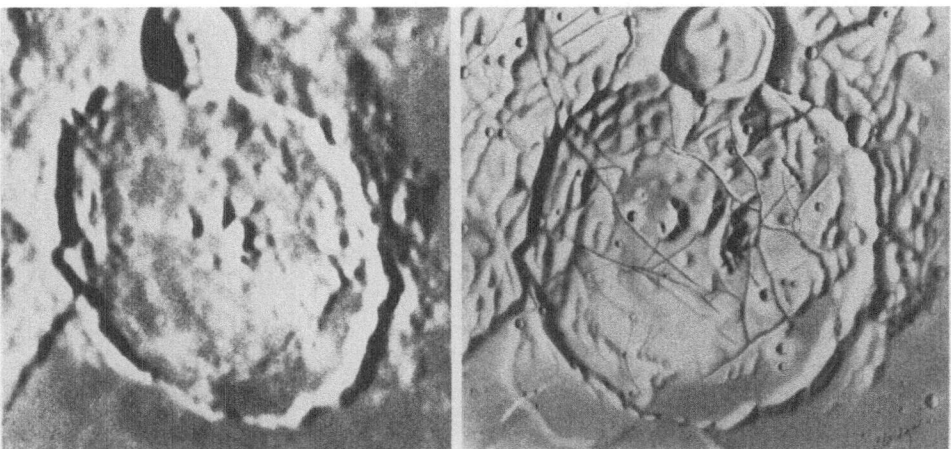

Fig. 7.3. On the left is a telescopic photograph of the lunar crater Gassendi. On the right is an ACIC drawing of the same area, rectified to scale, and including additional detail obtained by visual telescopic observations.

second and pick out the instant detail that is visible which is entirely lost in the photograph. That is, the eye sees detail by glimpses when 'seeing' is perfect and can ignore those instances of turbulence when the image is blurred; this, the photographic plate cannot do."

2. ACIC at Lowell Observatory

ACIC made several trips to Yerkes in the summer of 1960 to observe the Moon but it soon became evident that the number of clear nights at this location would not provide sufficient viewing time to support an extensive lunar mapping program. Also, Kuiper was leaving Yerkes to direct the Lunar and Planetary Laboratory at the University of Arizona. Before he left he advised ACIC to investigate some of the observatories in southwestern United States because in this area the average number of cloudy nights is between 40 and 80 a year as compared to 140 to 160 nights at Yerkes.

ACIC's first opportunity to view the Moon in the Southwest came through the courtesy of Arthur Hoag, director of the U.S. Naval Observatory's Flagstaff Station. It was in October 1960 when ACIC cartographer William Cannell visited the Navy's Flagstaff Station and was permitted to use their 40-in. reflector to observe the Kepler region. The seeing conditions were excellent and every night was crystal clear.

Impressive as the Navy's telescope was, it was impossible to consider using it for lunar mapping on a continuing basis because it was heavily scheduled with stellar work. However, Hoag suggested that Cannell approach John Hall, director of the nearby Lowell Observatory. This was done and when Hall saw the lunar work being accomplished by ACIC he agreed that their 24-in. refractor would be ideally suited for these observations since Percival Lowell, the founder of Lowell Observatory, had this telescope designed especially for visual observations of Mars and the planets.

Fortunately the Lowell 24-in. refractor was not in constant use and arrangements were made for Cannell to use this telescope for several nights. The resolution of small lunar features was impressive and seemed to be at least as good as the 40-in. Yerkes telescope. That was the beginning of the ACIC work at Lowell Observatory which amounted to only a few days once a month for the next year.

In the fall of 1960, two lunar charts, LAC-57 Kepler and LAC-75 Letronne were placed in work at ACIC as pilot editions based on telescopic observations. Monthly trips were established for visits to Lowell so that visual observations could be made during the two or three days required for the terminator to cross the charting area. Usually the morning terminator was observed but occasionally the stay was extended two weeks to observe the same region with the evening Sun. Cannell made most of the observations; sometimes he was assisted by Higgins or Moore of the ACIC Lunar Branch in St. Louis. On one trip Cannell was accompanied by Mrs Patricia Bridges who, because of her intense interest in the Moon and her exceptional ability to interpret the lunar forms and render them with an airbrush, became ACIC's lunar illustrator. Her lunar renditions seemed to be free from interpretive style. They were like sharp lunar photographs and thus gave a convincing impression of realism to the shape and character of features – some of her artistry is displayed in Figure 7.4.

NASA was favorably impressed with the first 1:1 000 000 scale LACs made from visual observations at Lowell and requested ACIC to map the entire Apollo zone. This was to be a total of ten LACs covering the area from 50° E to 50° W and from 16° N to 16° S. This meant that continuous observations would be required to meet a planned completion date of April 1964. It was estimated that the job could be accomplished by two observers and one illustrator working full time at Lowell Observatory.

3. The ACIC Lowell Office

ACIC contracted with Lowell Observatory to provide observing time on their 24-in. refractor and office space to accommodate three persons. Thus, the ACIC Lowell office was officially established on September 1, 1961, and staffed by ACIC personnel William Cannell, James Greenacre and Patricia Bridges. Cannell was placed in charge of the office and tasked along with Greenacre to do the observing while Mrs Bridges was to render the relief drawings. The three made an excellent team and completed five LACs during the first 12 months of operation.

The ACIC Lowell office, shown in Figure 7.5, had a distinctive charm which belied the space-age pursuits of its three occupants. This old wood-frame building on the Lowell grounds had housed the machine shop, carpenter shop and lumber storage for over 50 yr (it was vacant because a new masonry shop building had just been completed). With some drywall covering its open studs, a linoleum over its wide plank flooring, a gas space heater and a closet fitted with new plumbing, ACIC

Fig. 7.4. Relief features on LAC charts are drawn with an airbrush. On this drawing, the crater Copernicus is being developed in the third dimension under the skillful hands of scientific illustrator, Mrs Patricia Bridges.   *USAF Photo*

acquired 300 ft² of office space. Its location was ideal; only 300 ft from the 24-in. telescope.

A telephone link was established between the telescope and the office. Often, when Cannell or Greenacre was observing, Mrs Bridges would be working on the drawing in the office. When seeing conditions became exceptionally good or the illumination of some feature was of particular interest, Mrs Bridges was telephoned to hustle up to the telescope for a first-hand look as shown in Figure 7.6.

Visual observations of the Moon were normally made along the terminator and up to approx 30 deg in front of the terminator. In that region of the illuminated portion, the shadows are optimum for detail interpretation. Along the terminator, the very low and gentle relief features such as maria, ridges and valleys show up prominently. Craters, hills and rilles can be easily interpreted from 5 to 15 deg in front of the terminator. The very large or steep craters or mountains are best inter-

Fig. 7.5.   The ACIC Lowell office in 1961 was located in this wood-frame building, originally built in
early 1900's to house the Lowell Observatory machine and carpenter shop.   *Photo by R. Carder*

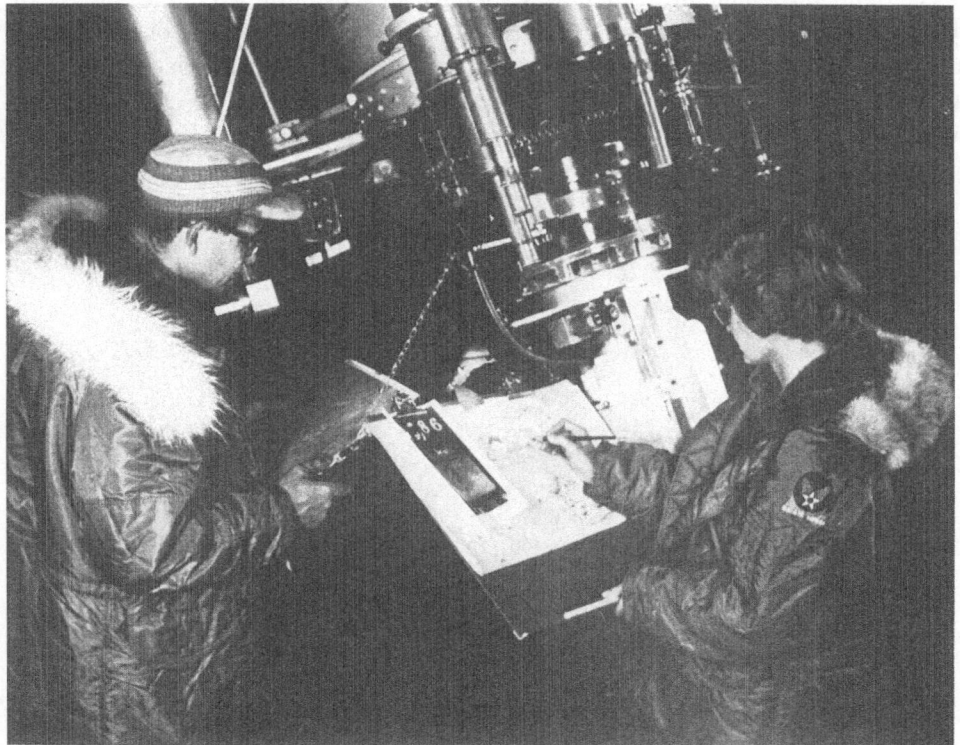

Fig. 7.6.   Visual telescopic observations played an important role in compiling the LAC charts. Here
William Cannell and Mrs Patricia Bridges are checking their work at the Lowell 24-in. refractor.

ACIC Photo

preted when the Sun angle is between 15 and 30 deg. Even higher illuminations can be important in seeing some crater floors and the fine details of crater rays.

The visual observer attempts to record his observations at the telescope by drawing sketches of what he sees or by making notes or annotations on a photograph of a region. During mediocre seeing, this may be a slow task because the observer must concentrate on a single feature and wait for the moments of steadiness to occur. When they occur frequently (every few seconds), the observer can make steady progress, but when the image is steady with only slow pulsations or swimming motion, the amount of small, fine details that can be seen at one time will usually overwhelm him so that he can do nothing but stare in amazement. At times like this the camera is a most valuable tool.

4. The Motion Picture Camera

It was John Hall, the director of Lowell Observatory, who suggested that a motion picture camera could be used to advantage to supplement visual observations because a large number of photographs taken in rapid sequence would increase the likelihood of catching a portion of the image when it is nearly stationary.

A 35-mm motion picture camera was obtained and custom-fitted to the tailpiece of the 24-in. telescope (see Figure 7.7). Thus, the telescope was equipped for both visual and photographic observations. The eyepiece and camera were arranged as one unit, with reflex prisms diverting the optical beam to the eyepiece. The observer could switch from visual to photographic observing by pulling a plunger which retracted

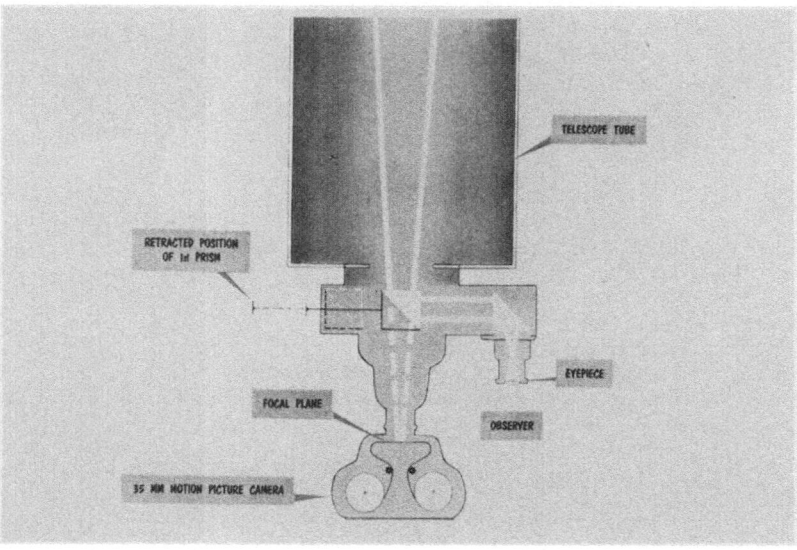

Fig. 7.7. This ACIC drawing illustrates the double reflex prism arrangement which allows for quick change-over from visual to photographic observations on the Lowell 24-in. telescope. When the first prism is retracted, light goes directly to the 35-mm motion picture camera at the prime focus.

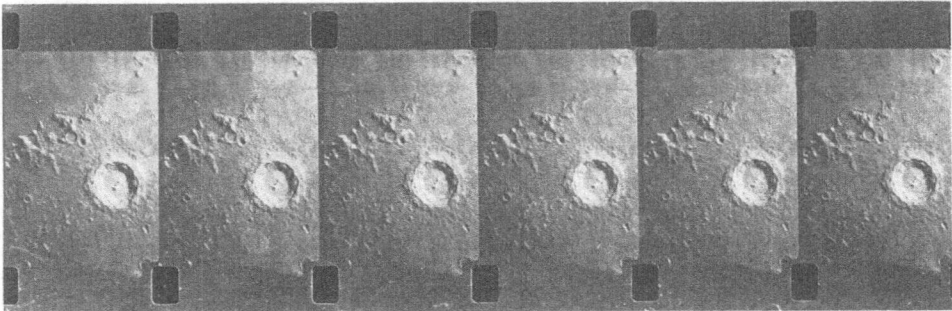

Fig. 7.8.   A series of 35-mm exposures taken with a movie camera attached to the Lowell 24-in. refractor. Careful examination will show sharp images on some frames that were exposed during moments of steady 'seeing'.   *ACIC Photo*

the first prism and allowed the beam to fall unreflected on the film. The second reflection to the eyepiece is introduced to erect the image inverted by the first reflection. This allows the second prism to be rotated about the optical axis so that the observer can assume a comfortable observing posture. A reasonable amount of comfort is important when the observer is expected to make concentrated visual observations for several hours.

Shown in Figures 7.8 and 7.9 are examples of lunar exposures taken in sequence with a movie camera. The best image is obtained when the exposure is made precisely during the instant the image is absolutely quiet. This rarely occurs, but by running the camera for a few seconds to obtain a burst of frames, the odds are favorable that some of the frames will record sharp images during a steady moment. Normally some images are taken on each night of observations. When processed the next day they serve to assist the observer because having clearly seen the same area in the telescope he can readily interpret his visual annotations recorded the night before. Later the 35-mm camera was replaced with a 70-mm Hulcher camera which provides a much larger field of view.

Hundreds of feet of film were being taken which soon developed into a logistics problem. Greenacre, having some prior photographing experience became the film and print processor. However, he could not cope with processing and printing large quantities of film in a 3 × 4-ft corner of the bathroom.

The photo lab problem was solved when ACIC authorized a new contract with Lowell Observatory to enlarge the office space and build a darkroom with a light trap entry, adequate sinks, counter space and storage for paper and film. The office space was also increased by fixing up more of the old carpenter shop and part of the old machine shop. By the fall of 1962, the ACIC Lowell office had acquired 700 ft² of office space and excellent photo lab facilities.

5. LAC Production

To keep pace with the rapidly expanding lunar charting workload, ACIC transferred

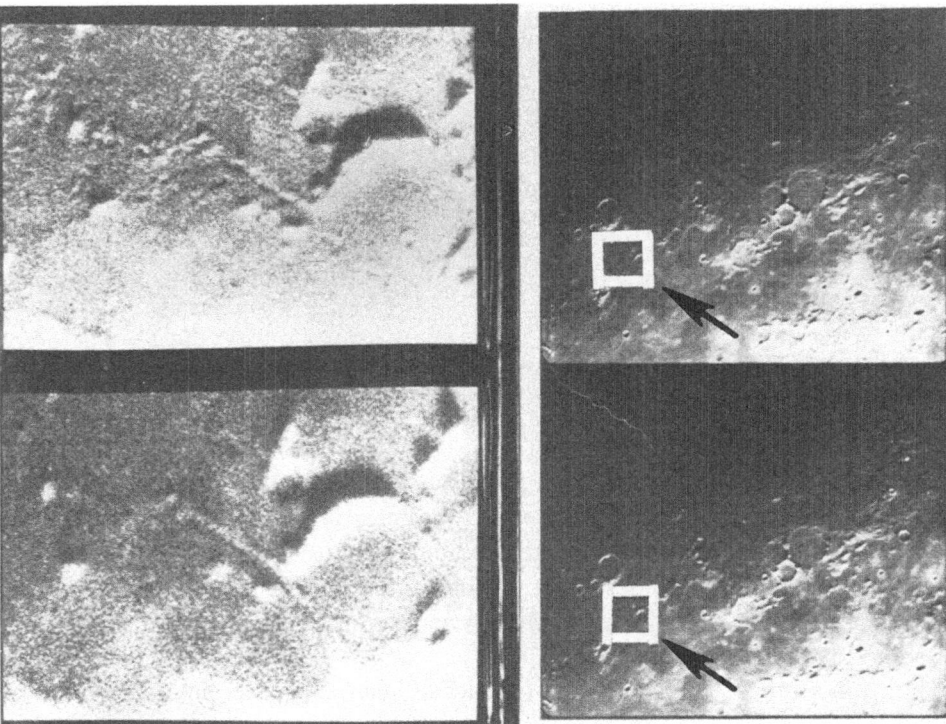

Fig. 7.9.   On the left are enlargements of identical areas from adjacent frames (right) taken with a movie
camera. Note how atmospheric turbulence can degrade the image in a fraction of a second.
ACIC Photo

two more cartographers, Leonard Martin and Fred Dungan, to Lowell. With the
help of Dungan's airbrush, as shown in Figure 7.10, ACIC was rapidly reaching a rate
of one LAC every six weeks. As the drawing for each LAC was finished, Cannell or
Greenacre personally transported it from Flagstaff to the University of Arizona in
Tucson for consultation with Kuiper, Arthur and Whitaker (shown in Figure 7.11
is Ewen Whitaker reviewing one of the LAC drawings). This team of lunar experts,
being intimately familiar with the Moon offered constructive suggestions on feature
portrayal – a procedure that ACIC followed for each lunar chart drawn from telescopic observations.

Six of the original ten LACs had been completed by the beginning of 1963, but
NASA had decided by this time to extend the LAC coverage as far as visual observations would permit. Obviously, limb charts were not feasible because foreshortening
was too severe, thus causing positioning and character of features to be very uncertain.
While it was feasible to compile 44 LACs from visual observations, the small ACIC
Lowell staff of five persons was insufficient to accomplish the job before the scheduled
flight of Apollo.

Rather than send more personnel from ACIC in St. Louis, it was decided to contract

Fig. 7.10.   Fred Dungan, an ACIC cartographer, skilled in using the airbrush, is working on one of the
LAC drawings.   *USAF Photo*

with Lowell Observatory for increased personnel. Two geographers from the University of Washington were hired in July 1963. The two new selenographers, Terrence McCann and Edward Barr immediately began learning to observe the Moon with the 24-in. refractor. McCann was teamed with Cannell to observe the region near the central meridian and Barr worked with Greenacre in the western limb area.

6. Lunar Color Phenomena

During the early evening of October 29, 1963, Greenacre and Barr were observing the Aristarchus region (LAC-39) when at 6:50 p.m. Mountain Standard Time (MST) Greenacre observed an unexpected color phenomenon on two prominences at the Cobra Head portion of Schröter's Valley and inside the southwestern wall of Aristar-

Fig. 7.11. Ewen Whitaker at the Lunar and Planetary Laboratory, University of Arizona, reviewing
one of the completed LAC drawings.   *Photo by R. Carder*

chus (see Figure 7.12, areas 1, 2 and 3). Greenacre immediately called Barr to share
this observation with him. Barr's first impression of the color was a dark orange.

The next step was to remove the No. 15 Wratten filter. Without the filter, the color
remained the same, but was brighter and had more sparkle. Both Greenacre and Barr
agreed that the color without the filter was reddish-orange.

The field of view was large enough to have all three color phenomenon areas in
sight at the same time. At approximately 7:00 p.m. MST, Greenacre noticed that the
spot of color at the Cobra Head (area 1, Figure 7.12) and on the hill across the valley
(area 2, Figure 7.12) had changed to a light ruby red, yet their density and sparkle was
sufficient to block out the surface underneath. He had the impression that he was
looking into a large polished ruby but could not see through it. The color phenomenon
was observed for 20 min before fading. (Greenacre and Barr are shown in Figure 7.13
standing beside the eyepiece of the 24-in. telescope).

A similar event was witnessed almost a lunar month later on November 27, 1963.
Again Barr was observing the Aristarchus region using the 24-in. refractor. At 5:30
p.m. MST, a streak of pink began to appear on the southwest exterior rim of Aristarchus (area 4 in Figure 7.12). Within a minute or two the color grew in intensity to a
brighter pink or light red.

Barr, satisfied that there was another probable color phenomenon taking place, immediately telephoned other observers in the ACIC Lowell office. The first to arrive was Fred Dungan, who had many hours of telescopic observing. Dungan had no difficulty in seeing the color and agreed with Barr that it was pinkish-red.

Greenacre arrived shortly thereafter and was able to confirm the observations of Barr and Dungan. He also called the Lowell director, John Hall, whose residence is only a short distance from the 24-in. dome. Hall arrived at the telescope in less than five minutes and had no trouble in seeing and verifying the color. As a matter of personal satisfaction and as a further check on his observing, he pinpointed the area on a photograph and a copy of the LAC-39. The three other observers agreed on the position.

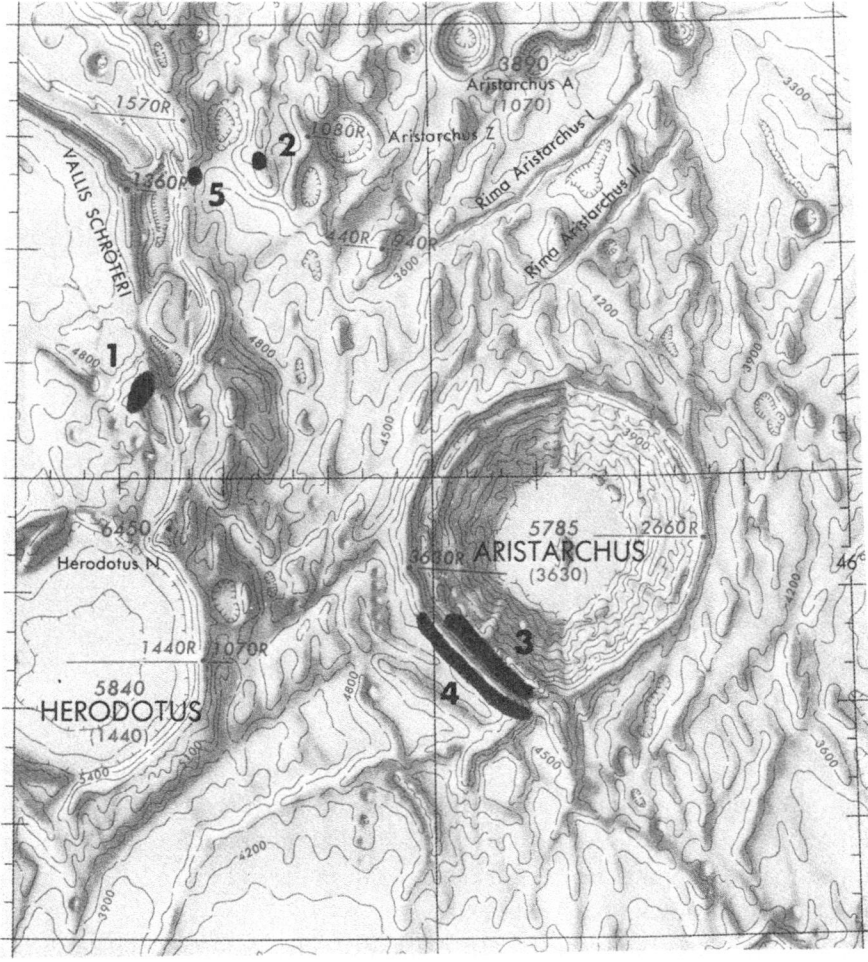

Fig. 7.12. Aristarchus, Herodotus and a portion of Schröter's Valley, from LAC 39, showing the locations of five color phenomenon observed at Lowell Observatory in 1963. Areas are numbered in order of appearance.

A short time after Hall had made his confirming observation, he telephoned to the nearby Perkins 69-in. reflector. He gave Peter Boyce, the observer on duty, a description of the color phenomenon and requested that Boyce try to make a spectroscopic scan of the area.

Unfortunately, Boyce had other instruments on the telescope and explained it

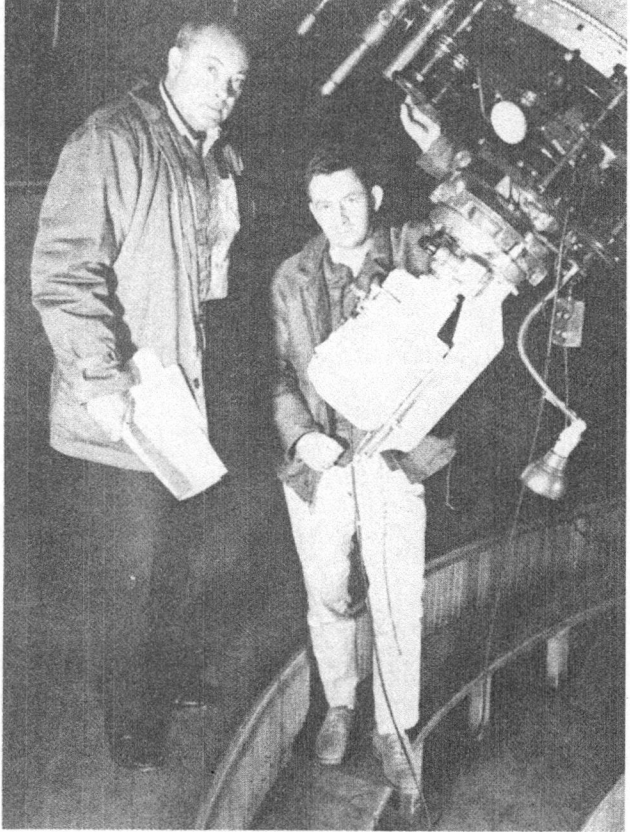

Fig. 7.13.   James Greenacre (left) and Edward Barr standing beside the eyepiece of the Lowell Observatory 24-in. refractor. Attached to the telescope is the 70-mm Hulcher movie camera.

Photo by W. Cannell

would take quite some time to remove them. Hall then suggested that Boyce make visual observations of the Aristarchus rim and adjacent areas. Boyce readily agreed and within fifteen minutes he called back to confirm the sighting of the reddish-pink glow on the southwest rim of Aristarchus.

Shortly after 6:09 p.m., the Lowell observers detected another reddish-orange spot on a hill to the east of Schröter's Valley (area 5 in Figure 7.12). Visual observations were continued until 6:30 p.m. when it became evident to all four observers that the color was rapidly fading.

Attempts by Greenacre to photograph the second color phenomenon with black and white Panatomic X film using the 70-mm Hulcher camera were unsuccessful. Neither were they able to detect the color phenomenon with the 12-in. refractor finder telescope attached to the 24-in. telescope. (This probably explains the absence of similar observations by amateurs with smaller telescopes).

The scientific significance of these two ACIC observations is that they seemed to be evidence of activity on the Moon; they also lent weight to the credibility of similar observations reported earlier. For example, Sir William Herschel in 1783 reported what he believed to be an erupting volcano in the Aristarchus region.

The ACIC color phenomenon observations were of particular interest to the U.S. Geological Survey selenologist who had located in Flagstaff to be near the supposed Earth analog of lunar craters – Meteor Crater. Certainly it tended to give evidence that the Moon was not a dead satellite as so many scientists had believed. However, it was not ACIC's concern to explain the color phenomena but rather to get on with the task of mapping the Moon.

7. The 20″ Morgan Telescope

It soon became evident that ACIC needed additional telescopic facilities if schedules were to be met, but Lowell had no more telescopes to rent. Fortunately, John Hall knew a Texas oil man and amateur astronomer, Benjamin Morgan, who owned a 20-in. Tinsley apochromatic refractor and was willing to sell it. After some tentative agreements were reached with ACIC, Roger Putnam, the trustee of Lowell, agreed to purchase the Morgan 20-in. telescope, move it to Flagstaff and rent it to ACIC.

The Morgan 20-in. telescope shown in Figure 7.14, arrived in early 1964; a new dome was built (see Figure 7.15) and it was dedicated for lunar work on April 15, 1964. Afterwards, the staff of the ACIC Lowell office was further enlarged and a new building was designed to accommodate as many as 20 persons.

8. ACIC Lowell Office Expands

A new building for the ACIC Lowell office, shown in Figure 7.16, was occupied in December 1964. Its appearance was more compatible with the space-age work going on inside. It is a low structure attached to the front of the old ACIC Lowell office to blend with it. The two buildings occupy about 3000 ft^2. The illustrators especially appreciated their new quarters with north-facing glass walls which permitted a high level of light without it ever being in direct sunlight.

Lowell hired a full-time photographer, Robert Maulfair, to assist ACIC. He took over all photographic processing and printing that had formerly been accomplished on a part-time basis by the observers. Maulfair arrived in time to be of great assistance on the Ranger VII charts.

The ACIC Lowell Group were intensely interested in the first Ranger photographs since for the first time they were showing very clearly the 200- to 300-m lunar features that they had been trying so hard to see with the telescope. The five Ranger charts (RLC 1–5) were assigned to Lowell for rendering and were accomplished by Mrs Bridges and Dungan in about 6 weeks.

ACIC tried to observe the actual impact of Ranger VII by stationing observers at

Fig. 7.14.   The Morgan 20-in. triplet refracting telescope. The tube is made of stainless steel. The $f/16$ lens is highly color corrected. Attached to the 20-in. is a 2-in. finder telescope, a 4-in. finder telescope (also a color corrected triplet) and a 10-in. guide telescope.   *Lowell Photo*

the 20-in. and 24-in. telescopes and the Lowell 69-in. Perkins reflector, expecting that a large plume of ejecta would be visible even if a sudden flash were not. Others across the country were doing the same thing but no one reported a sighting. The same kind of effort to observe the impact of Ranger VIII and IX failed to produce a sighting.

Fig. 7.15. The dome for the 20-in. Morgan telescope under construction. The rotating dome is made entirely of aluminum. It was officially turned over to ACIC for lunar observations on April 15, 1964.
Lowell Photo

Fig. 7.16. The new ACIC Lowell office completed in December 1964. It was attached to the front of the old building shown in Figure 7.5. *Photo by R. Carder*

9. Long Exposure Navy Plates

The need for astrometric photographs to support the ACIC selenodetic program was being met by taking sequential, short exposure photographs with the Pic du Midi 24-in. coudé refractor. The 24-in. Lowell refractor could not be used for this program because the scale was too small. The full lunar disc in the Lowell telescope was only 3 in. in diam while in the Pic du Midi telescope it was nearly 7 in.

In 1964 it occurred to Cannell that the new Navy 61-in. astrometric telescope located at the Naval Observatory's Flagstaff Station would be ideally suited for this kind of photography. Therefore, through cooperation with the station's director, Arthur Hoag, Cannell found that ideal selenodetic photographs could be made with this $f/10$, 61-in. telescope by using long exposures (longer than 20 s) during full Moon, with Kodak's superfine-grain spectroscopic emulsion 649F. This approach was taken because the $8'' \times 10''$ camera normally attached to the 61-in. telescope could not make exposures shorter than about 10 s.

Test shots made on the 61-in. telescope proved that long exposures up to 60 s using the Kodak 649F emulsion were fantastically sharp, especially when one considers that seeing conditions (regardless of how good they are) cause excursions that blur the photographic image. Apparently the slowness of the 649F emulsion caused a differential reaction to the seeing excursions. For example, the position of small bright craters is recorded at the mean position of the randomly oscillated image (see Figure 7.17). There was no need to measure sequential photographs in this case since there was no differential distortion caused by seeing.

Success with this long-exposure approach technique for obtaining full-moon photographs can be attributed mostly to the Navy's 61-in. precision drive-system for both right ascension and declination. However, it would require from 30 min to an hour to adjust both drive rates by trial and error to exactly match the lunar rate. More time was required for this adjustment when the exposure time was expected to be around 45 s.

This cooperative program between ACIC and the Navy of taking a few full-moon photographs during each lunation started in 1964 and continued through 1967.

10. The End of Visual Observations

The ACIC Lowell group was kept busy in 1964 and 1965 with the LACs and a new program of twenty 1:500000 scale Apollo Intermediate Charts (AICs) covering the equatorial region (described in the previous chapter). By July 1966 the work force had grown to 18 persons, but the end of visual observations was in sight. Only seven of the 44 LACs remained to be completed and the Orbiter was getting ready to fly.

The first three Orbiters had no influence on the LAC visual observations because they were near the equator while the remaining LACs were in the polar regions. Landing site maps were compiled in St. Louis from the first three Orbiters and the Lowell illustrators were called upon to render them with an airbrush.

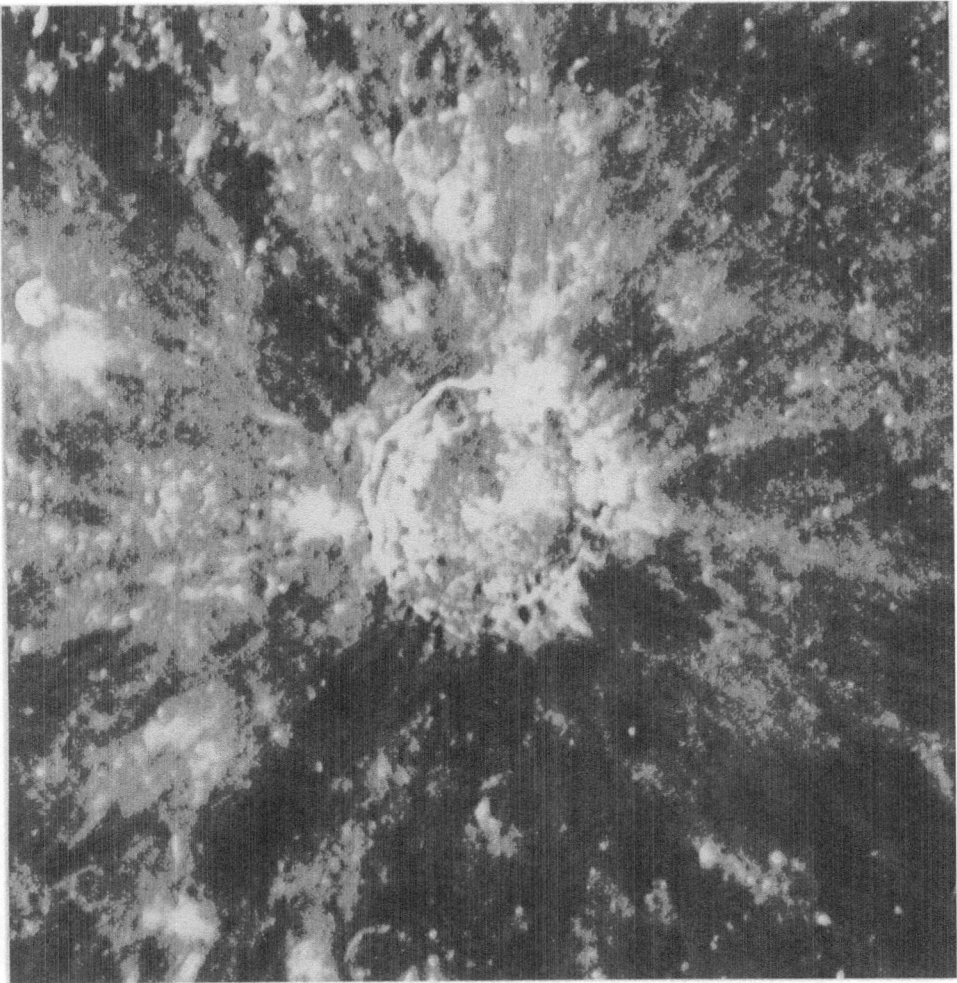

Fig. 7.17. A section of the Navy 61-in. Astrometric Plate No. 5818 showing the crater Copernicus and surrounding area. North is at the top. Universal Date and Time was October 9, 1965 – $06^h28^m43^s$. It was exposed 45 s on Kodak 649F emulsion with a GG 14 Filter. The small bright craters can be used as points for selenodetic measurements.   *Navy Photo*

The telescope was virtually obsolete with respect to lunar observations when Orbiter IV flew successfully and returned complete photographic coverage of the near side as well as some 75% of the far side. The Orbiter IV high-resolution photographs were used to finish the remaining few LAC by September 1967. A crash project to produce a new far side lunar map from Orbiters I, II, III, IV and Zond 3 photographs at a 1:5 000 000 scale was started in April 1967 and was completed in July in time for NASA to take it to the Prague IAU meeting in August 1967.

By 1968, eight illustrators had learned the Bridges technique of lunar illustration.

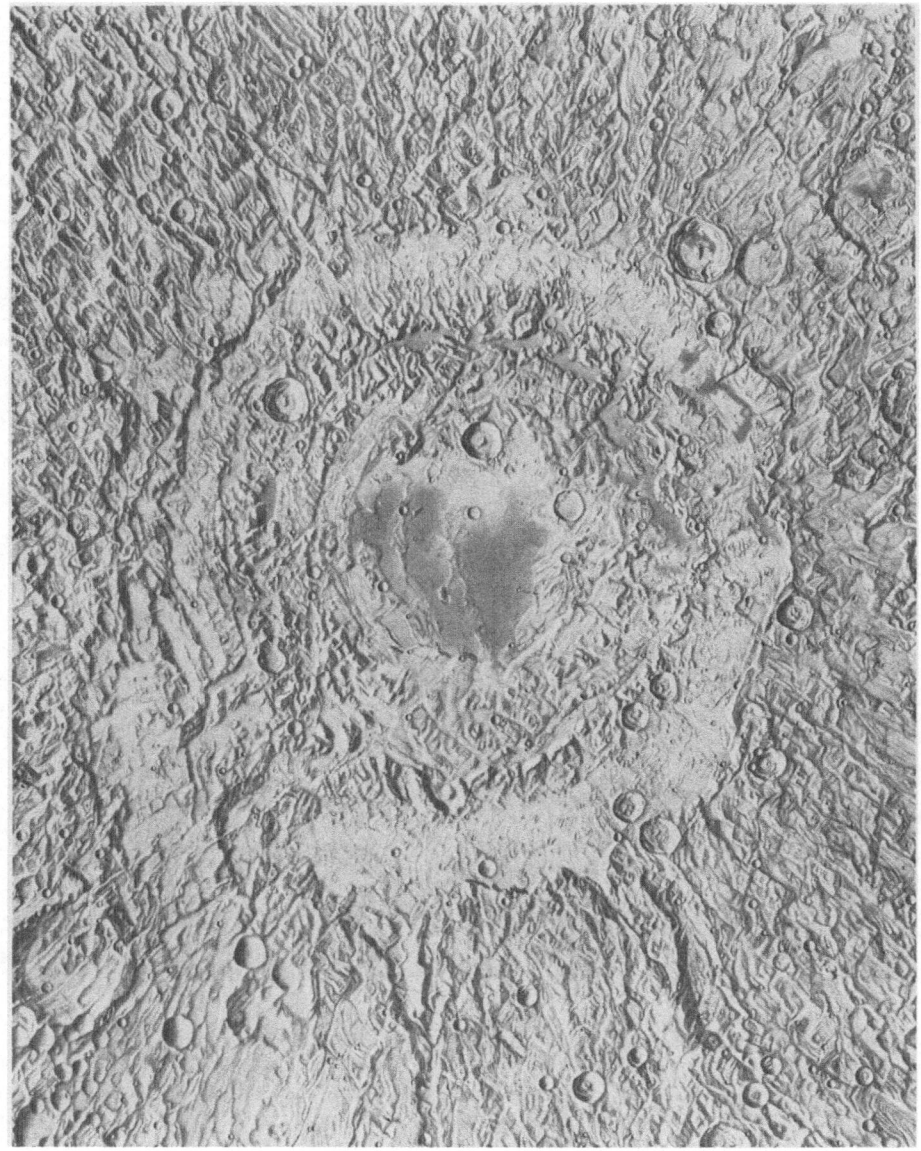

Fig. 7.18.   A shaded relief drawing of Mare Orientale produced by Mrs Patricia Bridges from Orbiter IV and V photographs. Mare Orientale, with an outer scarp 600 miles in diam, is centered at approximately 20° S lat. and 95° W long.

An outstanding example of her work from Orbiter photos is a drawing she made of Mare Orientale shown in Figure 7.18.

The drawings for the 16-in. NASA Lunar globe produced in late 1968 and early 1969 was the final lunar mapping efforts of the ACIC Lowell office. This was primarily the work of an outstanding new illustrator Jay Inge, who had taken the Bridges technique of lunar rendering and developed it even further.

The success of the ACIC team at Lowell cannot be measured by effort, patience and skill but it can be measured in terms of numbers of charts completed. Over an eight year period they compiled and drafted forty-four LACs and twenty AICs from telescopic observations. In addition, they rendered the drawings for seventeen Ranger charts, six Orbiter Site maps, two Surveyor Site maps and a lunar globe.

References

Greenacre, James A.: 1964, 'Lunar Color Phenomena', ACIC Technical Paper No. 12, May.
Holmes, Howard C. and Cannell, William D.: 1962, 'From Mars to the Moon', *Rev. Pop. Astron.*, September–October.

CHAPTER 8

U.S. AIR FORCE SPACE SUPPORT MAPPING

Early in 1959, the National Aeronautics and Space Administration (NASA) requested the Department of Defense (DOD) to provide cartographic support for the DOD forces that were to be utilized in recovery operations for Project Mercury. The USAF Aeronautical Chart and Information Center (ACIC) was identified as the agency most capable of providing this support because of experience gained in satisfactory similar requirements for aircraft operations.

1. Mercury

NASA's manned space-flight program started with Project Mercury, which aimed at placing an astronaut in orbital flight around the Earth, investigating his capabilities while in orbit and recovering him safely. One of the first cartographic items envisioned for support of the Mercury flights was an Earth orbital chart. To support this requirement, in May 1959 ACIC cartographer John Dornbach initiated plans to produce a Mercury Orbit Chart (MOC). A chart scale of approx 1:52000000 was dictated by the necessity to portray the Earth's circumference from 40° N–S lat on a Mercator projection within the 7-in. width of a chart holder. The holder was subsequently eliminated and the chart was cut into pages, along with note pages, check-off lists, etc., and inserted in a navigational aid book.

In conferences and discussions with representatives of NASA and McDonnell Aircraft (builders of the Mercury capsule) and particularly with several of the astronauts, general design criteria were formulated for the MOC. It was determined that the chart should be printed in colors that would approximate the natural appearance of major physiographical areas such as yellow-brown for the Sahara desert in northern Africa and blue-green for the jungles in Brazil. Consequently, the completed chart gave a kind of bird's eye view of the Earth.

As a result of opinions by the astronauts and others that shore line indentations and island shapes and patterns may be valuable checkpoints at the 100 to 120-mile altitude, ACIC produced an exceedingly precise drainage and island portrayal for a chart of this size. Its 15° graticule provided a satisfactory relationship to the rate of Earth's rotation and time. Criteria for selection of cities to be shown were based on such general factors as visibility to the astronaut by day, appearance at night due to artificial illumination, or strategic location in sparsely populated areas.

After the design problems in portraying base information had been resolved, the next task was to determine the best method of portraying all orbital and tracking data.

This information was to include successive orbit tracks around the Earth, elapsed time from launch along each orbit, retrorocket firing points, retrograde paths, re-entry and impact points, tracking station locations, recovery information and tele-metry and voice communication ranges of each station.

Several methods for portraying anywhere from three to nineteen complete orbits were investigated, including the use of colored lines, various types of line symbols and combinations of the two. Finally, in a more revolutionary approach, a line symbol was constructed of elongated type-style numerals which were placed in single file through their vertical axis to display the orbital track. Larger numerals were placed in boxes along the orbit track lines to present elapsed time from launch in hours and minutes at ten-minute intervals. The remaining portrayal of tracking sta-tions and recovery information, such as symbolization of the re-entry and impact points, posed a few problems. The resulting chart, MOC-3 shown in Figure 8.1, was used by astronaut John H. Glenn, Jr. in America's first manned orbital space-flight launched on February 20, 1962. A photograph of Astronaut Glenn holding the ACIC Mercury Orbit Chart is shown in Figure 8.2.

ACIC supported the operational requirements of the six manned Mercury flights consisting of the following additional cartographic products: a Mercury Earth Simulation Filmstrip, Mercury Test Flight Chart (MTC) and a Mercury Recovery Chart (MRC).

In October 1959, NASA established a requirement for a three-orbit 70-nm Earth-simulation filmstrip to generate Earth scenes for spacecraft attitude reference in a NASA simulator. The filmstrip image was projected onto a ten-foot plastic screen, which permitted the astronaut to view segments of the projected image through a periscope using the viewed image as an attitude reference. The art work provided only the general simulated appearance of the Earth's surface. The art work was con-trolled to a series of oblique Mercator projections with the mission ground track as the approximate line of tangency. An initial monochrome version of the filmstrip was produced and delivered to NASA in January 1960. The art work was subsequently printed on colored film and made available in May 1960.

The requirement for the MTC, at 1:500000, was established during 1960 to support the planning, control and recovery phases of a series of Mercury-Redstone (MR) 200 to 500-nautical mile down-range ballistic-trajectory test-launches from Cape Ca-naveral (subsequently renamed Cape Kennedy). The MTC was produced on an oblique Mercator projection with the 72° azimuth line from Cape Canaveral as the line of tangency.

Requirements for the Mercury Recovery Series were established in November 1960. This ten-sheet 1:5000000 scale series is centered on the equator, and covers 80° of lat and 360° of long. The primary purpose for the MRC series was to provide DOD sea/land/air recovery forces with a planning and plotting chart for spacecraft recovery operations.

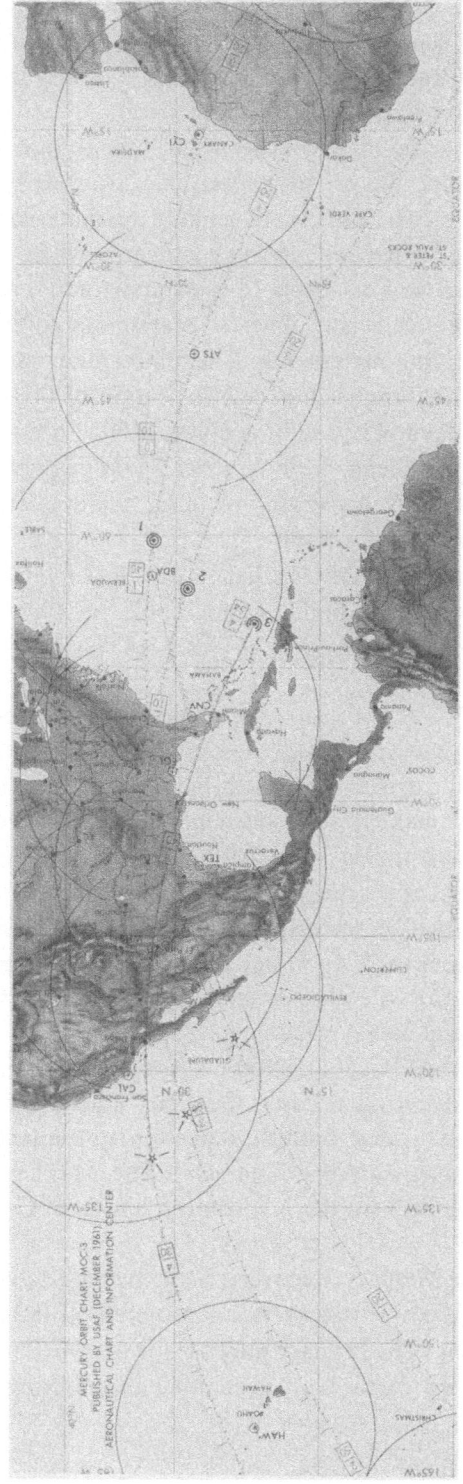

Fig. 8.1. A section of the Mercury Orbit Chart (MOC-3) used by astronaut John Glenn, Jr. in America's first manned orbital space-flight launched in February 1962. Bull's-eyes positioned on each orbit represents recovery zones.

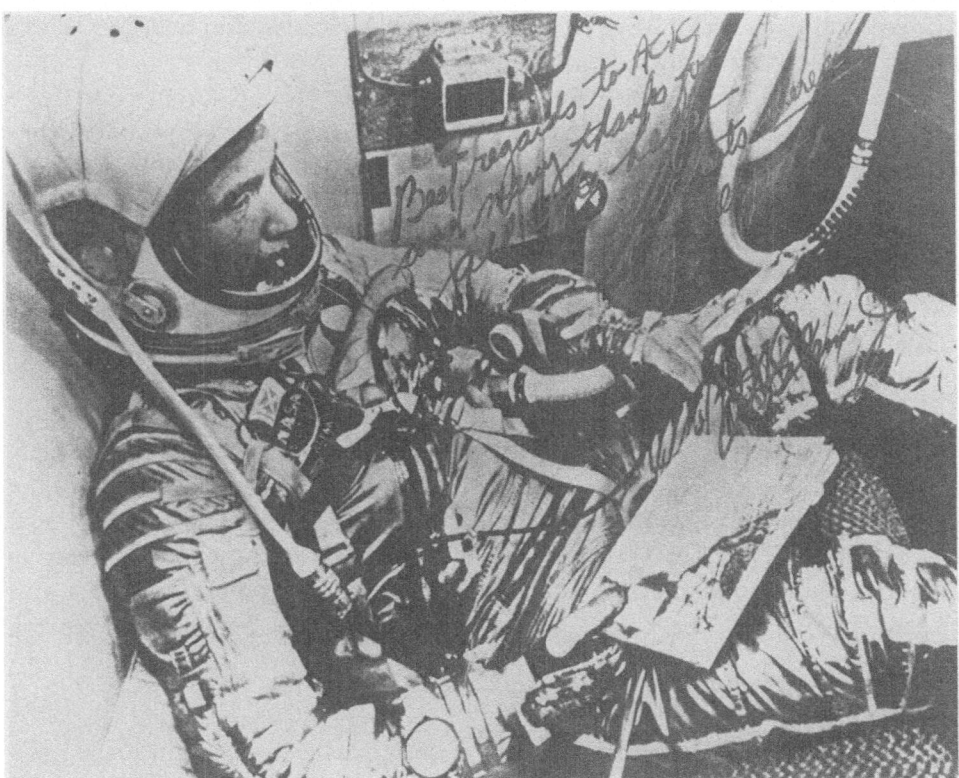

Fig. 8.2.   Astronaut John Glenn, Jr. in his space suit holding a copy of the ACIC produced Mercury
Orbit Chart (MOC-3).   *NASA Photo*

2. Gemini

Mercury was a necessary prelude to more extensive manned space-flights, and originally was to be followed by the Apollo program for sustained flight in Earth orbit and a trip around the Moon. But several factors prevented the direct jump from Mercury to Apollo. As Apollo studies progressed, the gaps in available knowledge became apparent. Mercury had proved that man could travel in space but it was not nearly enough experience to serve as the basis for a program of Apollo's magnitude.

There was no available expertise on rendezvous in orbit which was the key to the success of the Apollo program. Further, there was a need to determine man's ability to withstand the rigors of space environment and to develop all the techniques required for the lunar journey. The Gemini program grew out of those needs. It served to bridge the gap in technology and operational know-how between Mercury and Apollo, and it developed the skills and assembled the flight crews, support personnel and facilities that would be needed when Apollo reached operational status.

Twelve Gemini missions, the first two unmanned, were launched from May 1964 through November 1966. In support of these Gemini missions ACIC, initially under the technical guidance of ACIC personnel Joseph McKinney and later by Charles Miller, produced the following cartographic products:

Gemini Orbit Chart (GOC). Requirements for the Gemini Orbit Chart were identical to the requirements for the Mercury Orbit Chart. However, several modifications were made. The scale was changed from 1:52 000 000 to 1:60 000 000 to adapt the sheet size to the Gemini Plotting Board, and a one-degree graticule replaced the fifteen-degree graticule to reduce plotting time. The GOC, overprinted with selected Gemini mission data, was designed to satisfy the spacecraft crews' requirement for an in-orbit navigation aid. Ground track-data, contained on transparent plastic overlays, was registered to one basic chart. Using the chart and overlay, the astronauts could determine their nadir point on the Earth's surface and the communication range of the NASA and DOD tracking stations. Shown in Figure 8.3 is GOC-3 which supported the first Gemini manned flight launched in March 1965.

Gemini Mission Chart (GMC). The GMC is a 1:52 000 000 scale version of the GOC designed to satisfy NASA and DOD requirements for a small scale Operational and Planning Chart. The GMC was tailored to satisfy operational requirements unique to each Gemini mission. This data consisted of mission ground tracks with elapsed-time intervals, tracking stations and their capabilities, and planned recovery

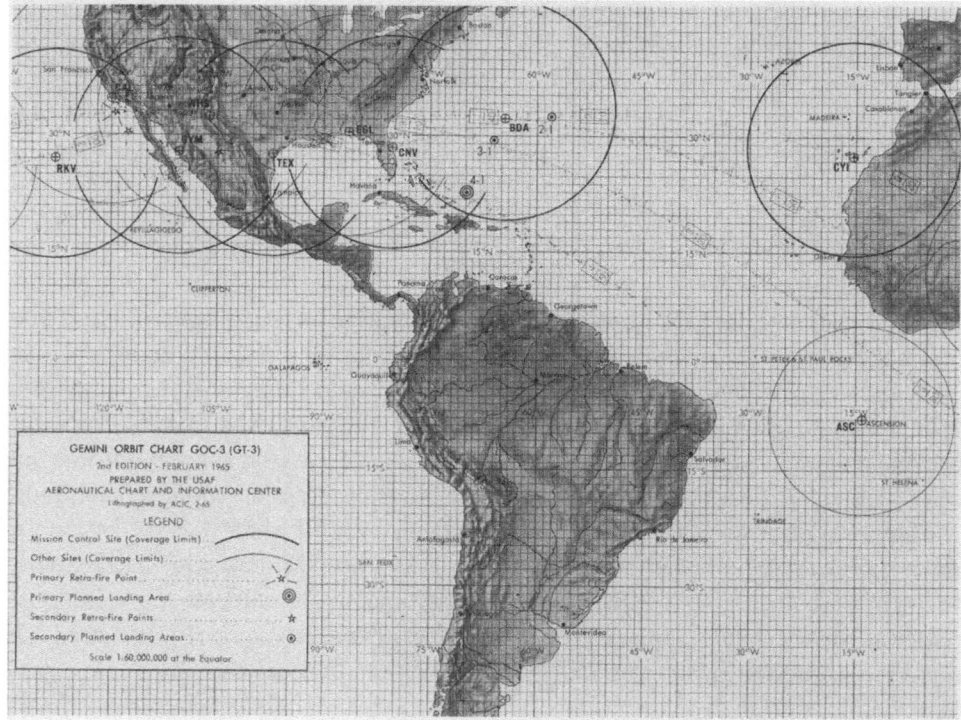

Fig. 8.3.   A section of the Gemini Orbit Chart (GOC-3) which supported the first Gemini manned flight launched in March 1965.

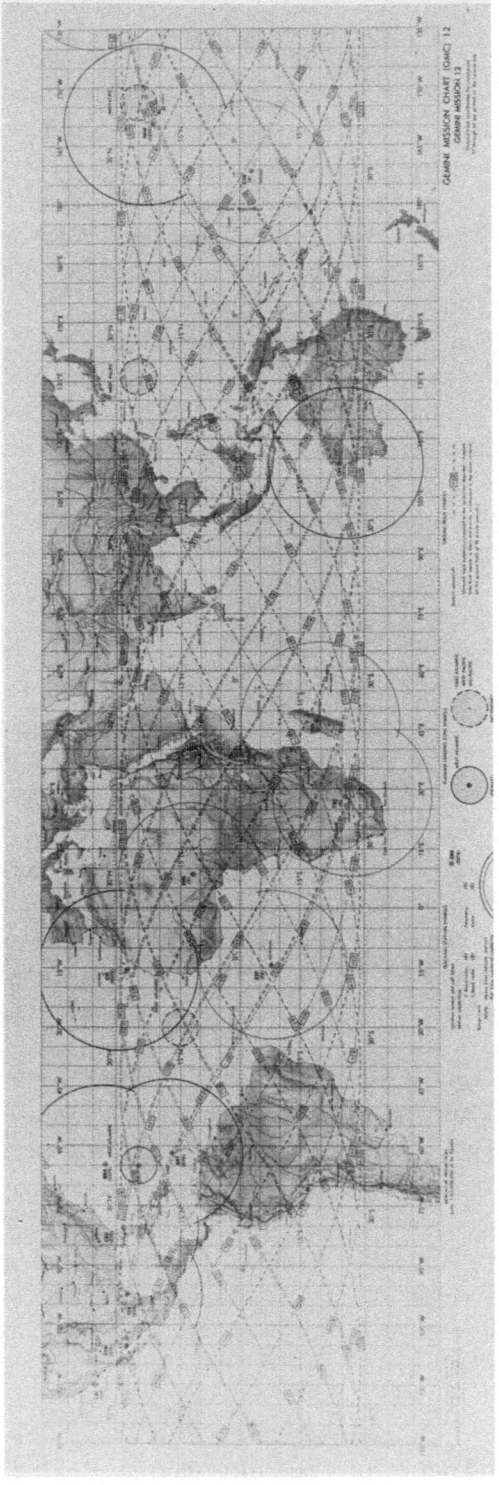

Fig. 8.4.   Gemini Mission Chart (GMC-12) used on the last Gemini flight launched in November 1966.

zones. Shown in Figure 8.4 is GMC-12 used on the last Gemini flight launched in November 1966.

Gemini Mission Plotting Chart (GMP). The GMP is identical to the GMC except that unstable mission data was omitted. The GMP was used as a plotting base to annotate supplementary information for planning operations unique to each major support function and to plot real-time data in the event of a change in the mission flight plan.

Spacecraft Weather Plotting Chart (NWC). The NWC is a monochrome version of the GOC tailored to plot weather data for use by spacecraft crews and mission control personnel before and during the mission.

Spacecraft Launch Abort Analysis Chart (NAA). The NAA at 1:12233000 scale on an Mercator projection covers an area from Eastern United States eastward to the African Coast. It was used to plan specific tasks to be performed by recovery crews in the event of an abort during the launch or early orbit phases of a mission.

Spacecraft Recovery Chart (NSR). The Mercury and Gemini requirements for a DOD Recovery Force plotting chart series were identical. For this reason, the 10-sheet 1:5000000 scale Mercury Recovery Chart series was used to produce recovery charts for Gemini.

3. Apollo

ACIC operational support of Apollo included the same types of graphics that were provided for Mercury and Gemini. In additional, new requirements were developed for charts and graphics to support the lunar mission operations including lunar orbit and landing phases. At ACIC this work was accomplished under the project direction of Robert Carder and Charles Miller.

The first Apollo flight, AS-203, launched in July 1966 was an unmanned sub-orbital flight designed to obtain flight characteristics of the Saturn launch vehicle. This was followed by eight flights before the now famous lunar landing on July 20, 1969 of Apollo 11. Shown in Figure 8.5 is a section of the Apollo 11 Lunar Module Descent Monitoring Chart annotated with the ground track descending to the landing area in Mare Tranquillitatis.

After Apollo 11, the United States astronauts returned to the Moon five times before concluding the Apollo program in December 1972. Shown in Figure 8.6 are the locations of the six Apollo landing areas. (Apollo 13 was a 'mission abort' because of service-module oxygen tank failure.)

In support of Apollo 17, the last lunar landing mission, ACIC produced an array of charts required for liftoff, Earth orbit, lunar orbit, lunar landing, lunar departure and Earth recovery. (A description of these graphics and how they were used for a 1486000-mile round trip to Taurus-Littrow is covered in the remaining part of this chapter.)

3.1. MISSION PROFILE

The Apollo Translunar/Transearth Trajectory Plotting Chart (ATT) serves as a

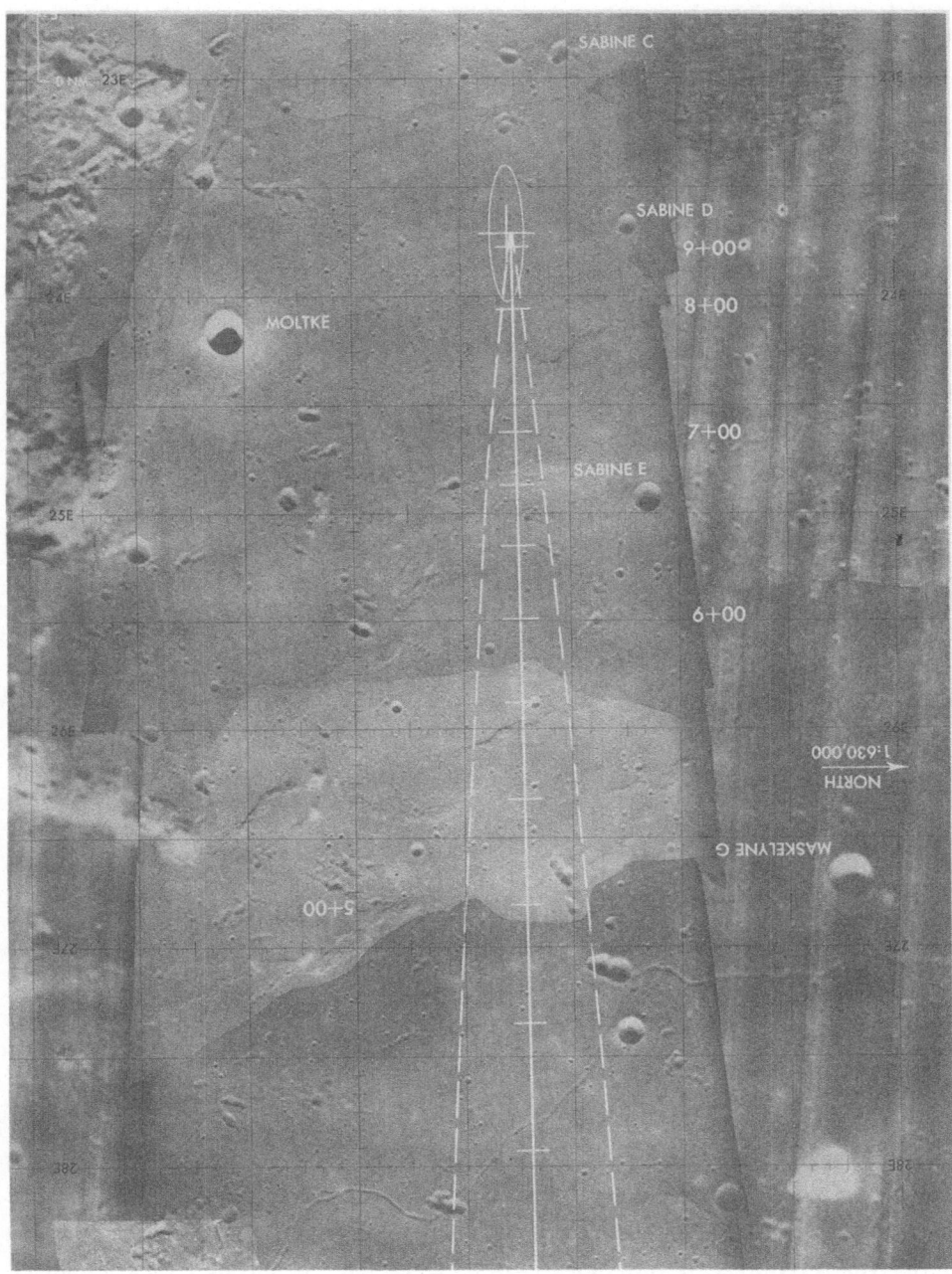

Fig. 8.5.   A section of the Apollo 11 Lunar Module Descent Monitoring Chart. The landing site in Mare Tranquillitatis is marked by an ellipse. Numbers beside the ground track indicate elapsed time in minutes from power descent initiation.

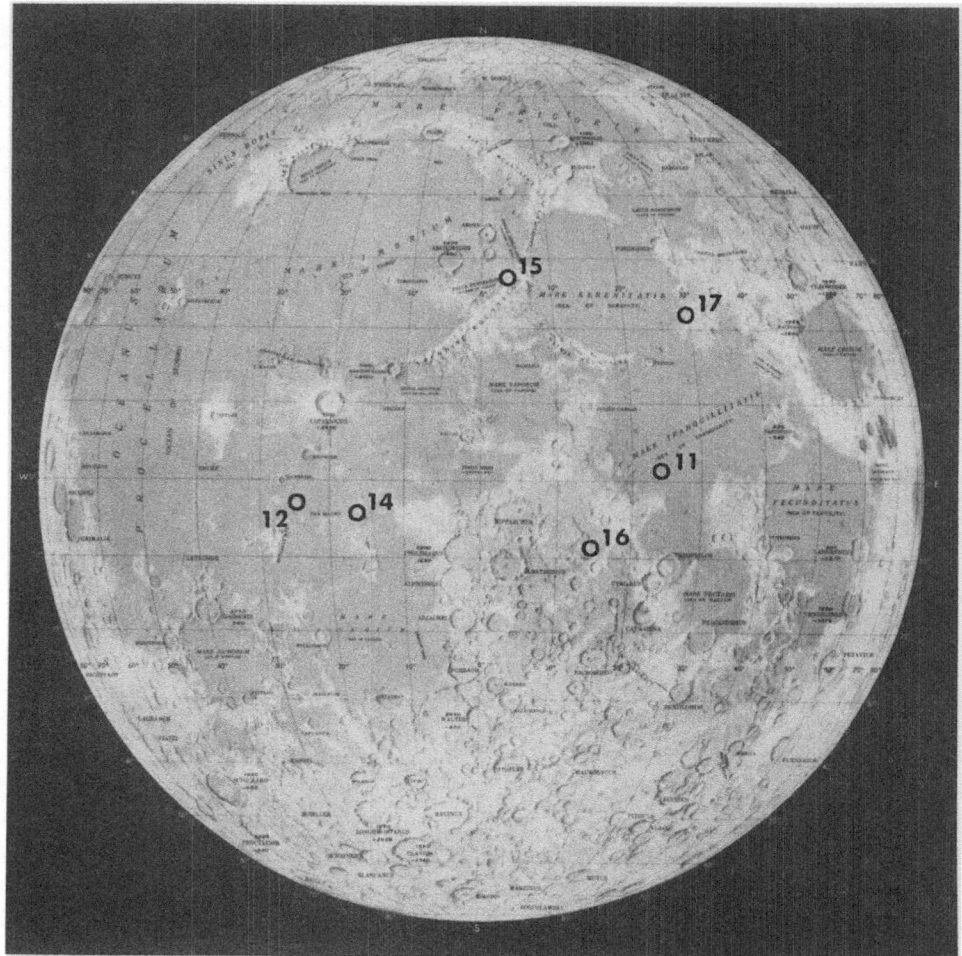

Fig. 8.6.   Circles mark the location of six Apollo manned landings on the Moon.

general reference chart for NASA mission-support components. This chart presents a polar view of mission profile and depicts major mission events from Earth launch through lunar landing and final return to Earth. The ATT chart, 20″ × 24″, shows concentric circles centered on the Earth at 50 000 nautical mile intervals with the outermost circle at 200 000 nm. Just beyond the 200 000-nm range, the daily position of the Moon is indicated from new Moon on December 5th through the last quarter on December 29, 1972.

In the lower right corner of the ATT chart is a general description of the mission from liftoff at Cape Kennedy to splashdown in the Pacific Ocean. In the lower left is a table of significant events of the Command Module (CM) and the Lunar Module (LM). These events are shown by Ground Elapsed Time from launch, Central Standard Time and date.

Figure 8.7 shows a section of the ATT chart. From this view, the mission profile can be traced from launch to earthparking orbit to translunar injection followed by a translunar coast to a rendezvous with the Moon three days later on December 10, 1972. At 1:49 p.m. CST on December 10th, the spacecraft performed an insertion

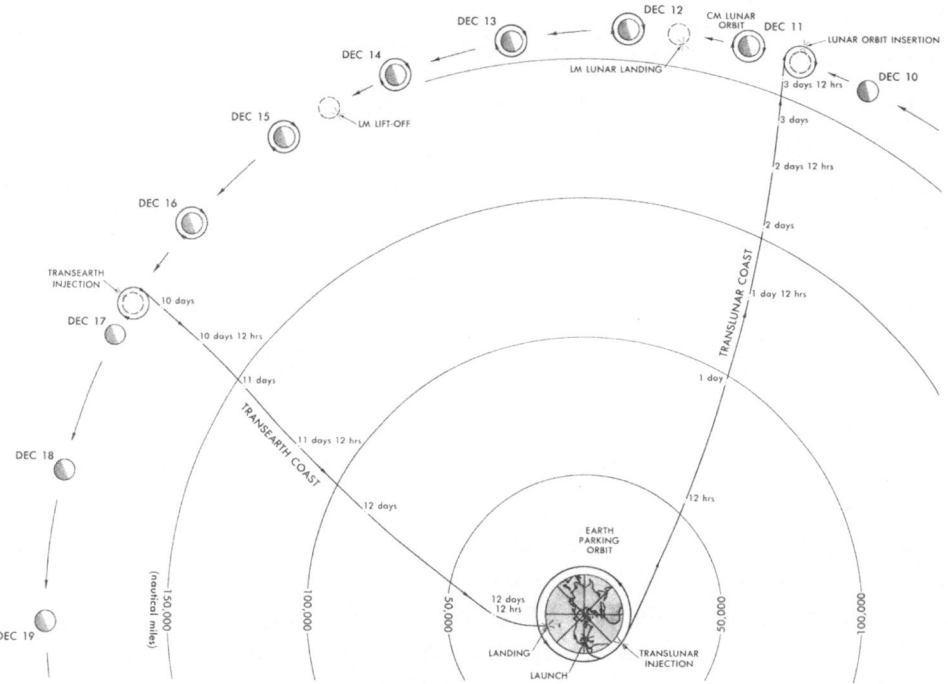

Fig. 8.7.   A section of the Apollo 17 Translunar/Transearth Trajectory Plotting Chart showing a polar view of the mission profile from Earth launch to the Moon and return.

burn to place it into a lunar orbit, shown in upper right of Figure 8.7. The dashed circular outlines depict the relative positions of the Moon when lunar orbit insertion, landing and liftoff occurred.

3.2. LIFTOFF

The Launch Complex Graphic (LCG), 26″ × 31″, compiled at a scale of 1:21 400 (approx 3 in. to the mile) covers the Apollo launch site at Cape Kennedy. This chart is used for monitoring launch activities and possible abort operations. Figure 8.8 shows a section of the LCG with a protractor centered on Pad 39A. Leading to the west from Pad 39A is the causeway to the Vehicle Assembly Building, a distance of 5.5 km. Cape Kennedy's dual launch pads (39A and 39B) were originally constructed to expedite launch turnaround time. However, all Apollo launches, with the exception of Apollo 10, were launched from 39A.

The mission planning consideration for the launch phase of a lunar mission are, to

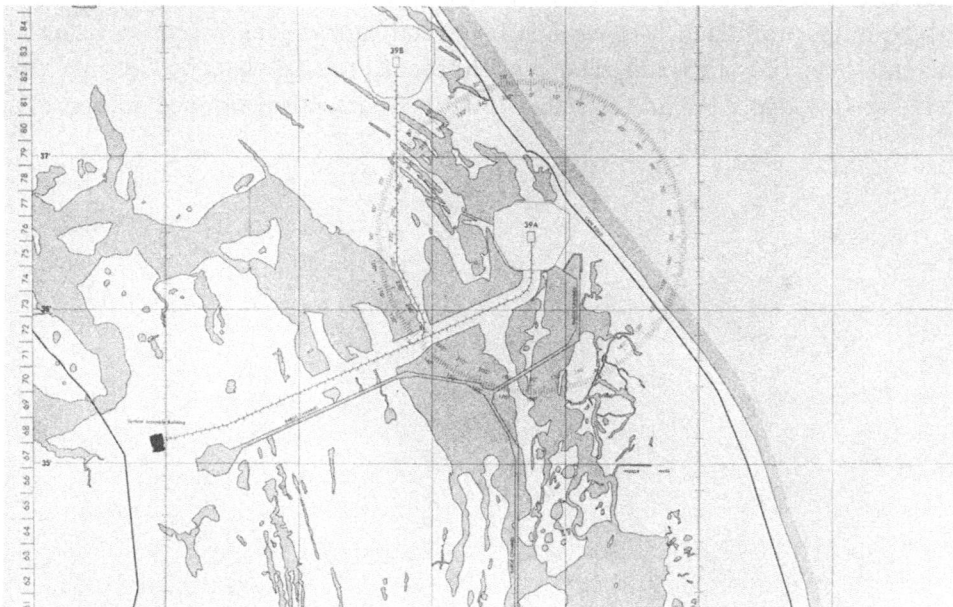

Fig. 8.8.   A section of the Launch Complex Graphic covering the Apollo launch site at Cape Kennedy.

a major extent, related to launch windows. During lunar landing operations, the Sun must be behind the Lunar Module and low on the horizon to accentuate small protuberances at the landing site. This requirement defines the 'monthly launch window' of a day or days which meet the mission operational constraints during a given month or lunar cycle. In addition, a 'daily launch window' (normally a few hours) exists which imposes further constraints. In December 1972 there were two days, December 6th and 7th, which would permit an ideal landing at Taurus-Littrow.

3.3. EARTH ORBIT

The Apollo Earth Orbit Chart (AEO) is a 1:40000000 scale chart of the Earth on a Mercator projection limited to 45° N–S lat, which was originally produced as the Mercury Orbit Chart.

The AEO chart is designed to provide information for monitoring the space-craft's flight from time of launch through initiation of translunar injection. Over-printed in light blue are ground tracks for the first revolution emanating from Cape Kennedy at launch azimuths of 72° through 100°, which are the flight azimuth limits dictated by safety (overflying populated areas) and launch window restrictions. Within these limits, the launch azimuth may be advanced to accommodate necessary changes in time of launch and still effect a desired orbit rendezvous over the Moon's antipode where translunar injection occurs. For example, Apollo 17 was scheduled to be launched at a 72° azimuth but after a delay of 2 hr and 40 min it was launched on an azimuth of 91 30'.

Figure 8.9 shows a section of the AEO chart covering the launch site and down-range over the South Atlantic. During the third revolution, the Saturn's Third Stage was ignited upon reaching the Moon's antipode, shown by dashed lines in Figure 8.9, just north of South America thereby injecting the spacecraft into a translunar tra-

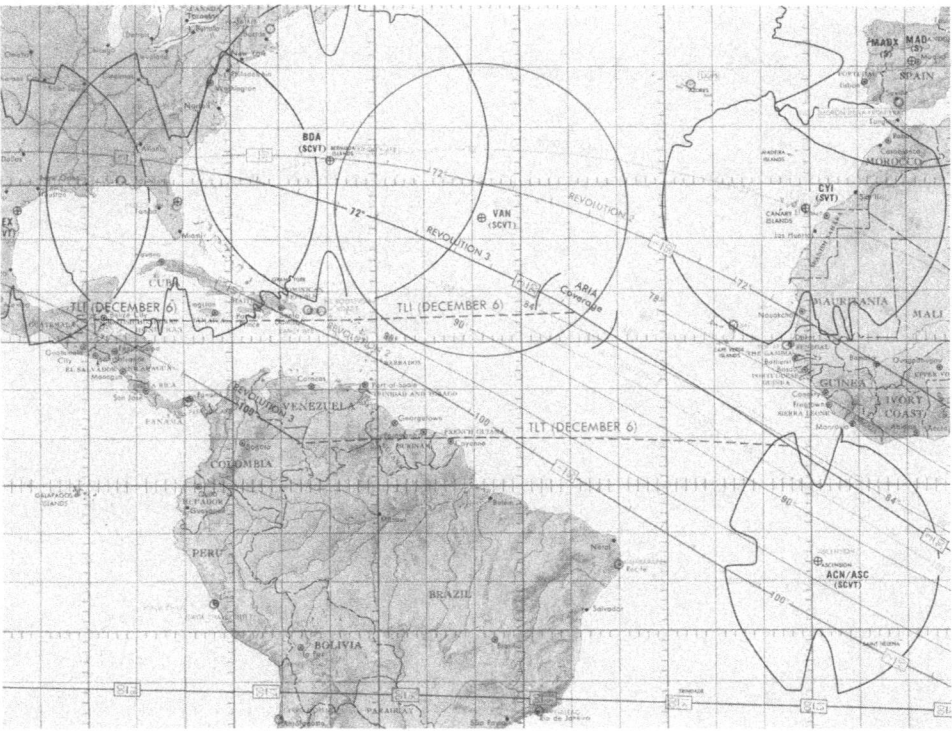

Fig. 8.9. A section of the Apollo 17 Earth Orbit Chart covering the launch site and downrange over the South Atlantic. Shown are the Manned Spaceflight Network tracking stations with their call letters and tracking limits. The station in the center of the circle is Vanguard, a radar tracking ship at sea.

jectory. Time in hours and minutes is designated at ten-minute intervals along the 72° and 100° launch azimuth ground tracks. Also indicated on Figure 8.9 are some of the Manned Space Flight Network tracking stations, along with their call letters and tracking limits.

3.4. LUNAR ORBIT AND LANDING

For monitoring the lunar orbital phase of the mission, ACIC produced the Apollo Lunar Orbit Chart (ALO), which is similar to the AEO chart of the Earth. The ALO was compiled on a Mercator projection at a scale of 1:11000000 with latitude coverage limited to 40° N–S. The ALO base shows the lunar surface in considerable detail since it was reduced from the original LOC drawings described in an earlier chapter.

Lunar Orbit Insertion (LOI) of Apollo 17 occurred on the Moon's far side in sunlight. The ground tracks for orbits 1, 10, 20, 30, 40 and 47 are shown in blue. On the 48th revolution a Lunar Orbit Plane Change (LOPC) was required to establish a coplaner orbit with the Lunar Module ascent trajectory and rendezvous. Following this the orbits 49, 60, 70 and 75 are printed in magenta.

Shown in Figure 8.10 is a section of the ALO chart covering the landing area. The

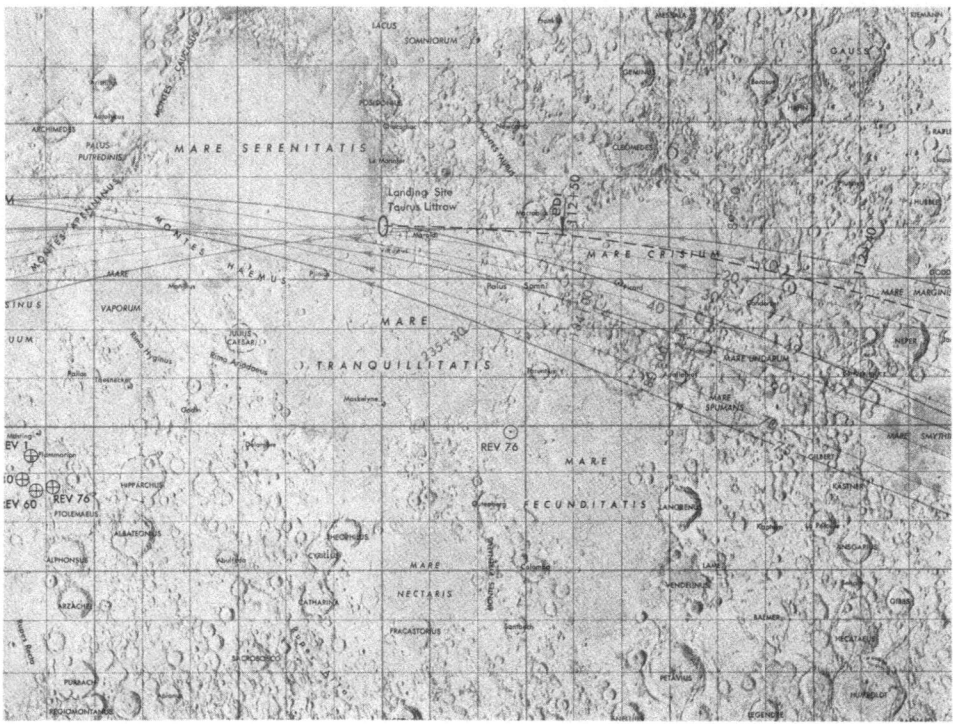

Fig. 8.10.   A section of the Apollo 17 Lunar Orbit Chart covering the Taurus-Littrow landing area. Shown are the ground tracks for orbits 1 through 75 at ten orbit intervals.

Taurus-Littrow landing site is designated by an ellipse and printed in red. Elapsed mission time from launch is shown in hours and minutes along each ground track while the number inside the line depicts the orbit revolution. The point along the ground track where powered descent is initiated is identified by an arrow and labeled PDI.

The Lunar Orbit Insertion burn placed Apollo 17 into a 60 nm × 170 nm elliptical orbit with perilune over the landing site. On the second lunar revolution, the Descent Orbit Insertion burn lowered the orbit to an 8 nm × 100 nm elliptical orbit, again with perilune over the landing site. For the next ten revolutions the Apollo crew used the CM/LM Orbit Monitor Chart (CDM).

The CDM chart, which covers the approach corridor and landing area, is used on board both the Command Module and Lunar Module to support the CM and LM crews, lunar observations at orbiting altitudes of less than 20 nautical miles. For Apollo 17, the photo imagery of the CDM strip chart provided 70° longitudinal coverage approaching the landing area and 15° beyond. Because of its length (approx 12 ft) it has to be printed in three sheets. For on board use the three sheets are cut into strips 10½ in. wide, centered on the last revolution before lunar landing (revolution 12), and spliced into a continuous strip, then accordian-folded with panels numbered 1

Fig. 8.11. The Apollo CM/LM Orbit Monitor Chart is used for monitoring the approach corridor and landing area at orbiting altitudes of less than 20 nautical miles. This 12-ft strip chart is accordian-folded for convenience in using on board the spacecraft. Chart shown was used for Apollo 16.

through 18 for reference. This provides a convenient book format strip-chart as shown in Figure 8.11.

The CDM chart portrays mission data such as the CM/LM 3rd and 12th revolution ground track. Elapsed time in minutes and seconds both to and from Taurus-Littrow landing site is shown in the left margin. Figure 8.12 shown panel 15 of the CDM chart. In the lower left, the figure 1 + 00 indicates 1 min and 00 s before passing over the landing area designated by an ellipse and numbered 17–1.

3.5. LUNAR DEPARTURE

For departing from the Moon, the LM crew used the Ascent Monitoring Chart (LMA) which is shown in Figure 8.13. This is a single sheet (2 panel) chart covering approx 150 nm of the LM ascent trajectory from the landing site. Plotted on the LMA is the launch track which has a different azimuth from that used for landing. To permit monitoring of the rate of ascent, the ground track is ticked in 30-s time intervals after liftoff and labeled at 1-min intervals in the right hand margin.

While in lunar orbit, specified photographic assignments were carried out with the Hasselblad camera from inside the Command Module. For accomplishing this part

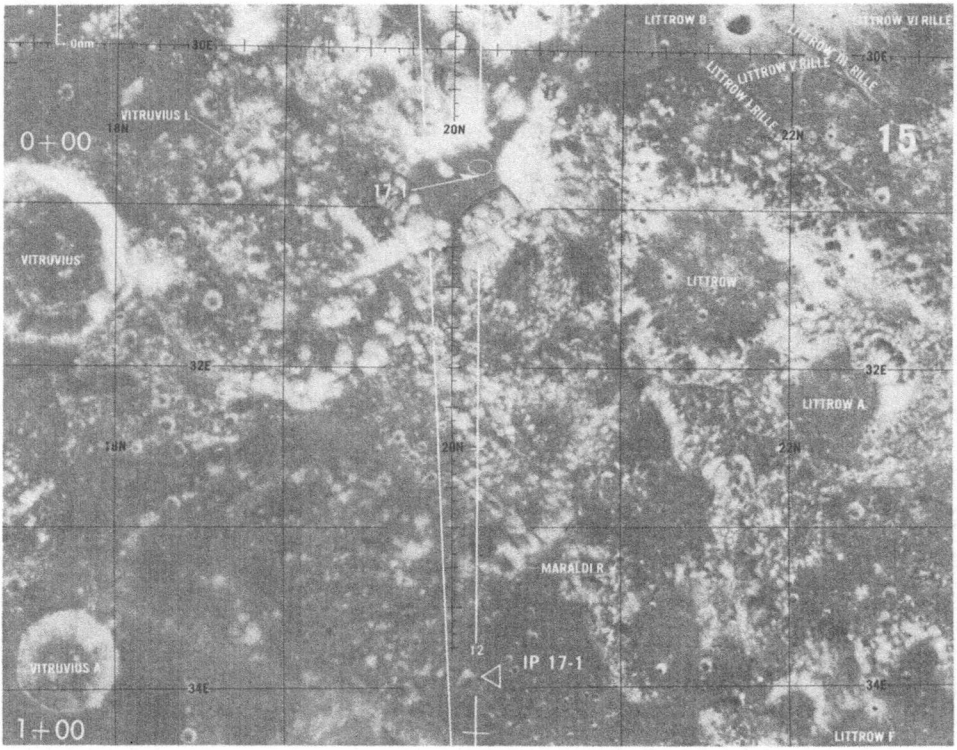

Fig. 8.12. This is panel 15 of the Apollo 17 CM/LM Orbit Monitor Chart. The Taurus-Littrow landing site is outlined by an ellipse and numbered 17–1. High Sun angle photographs were used in the compilation of this photomap.

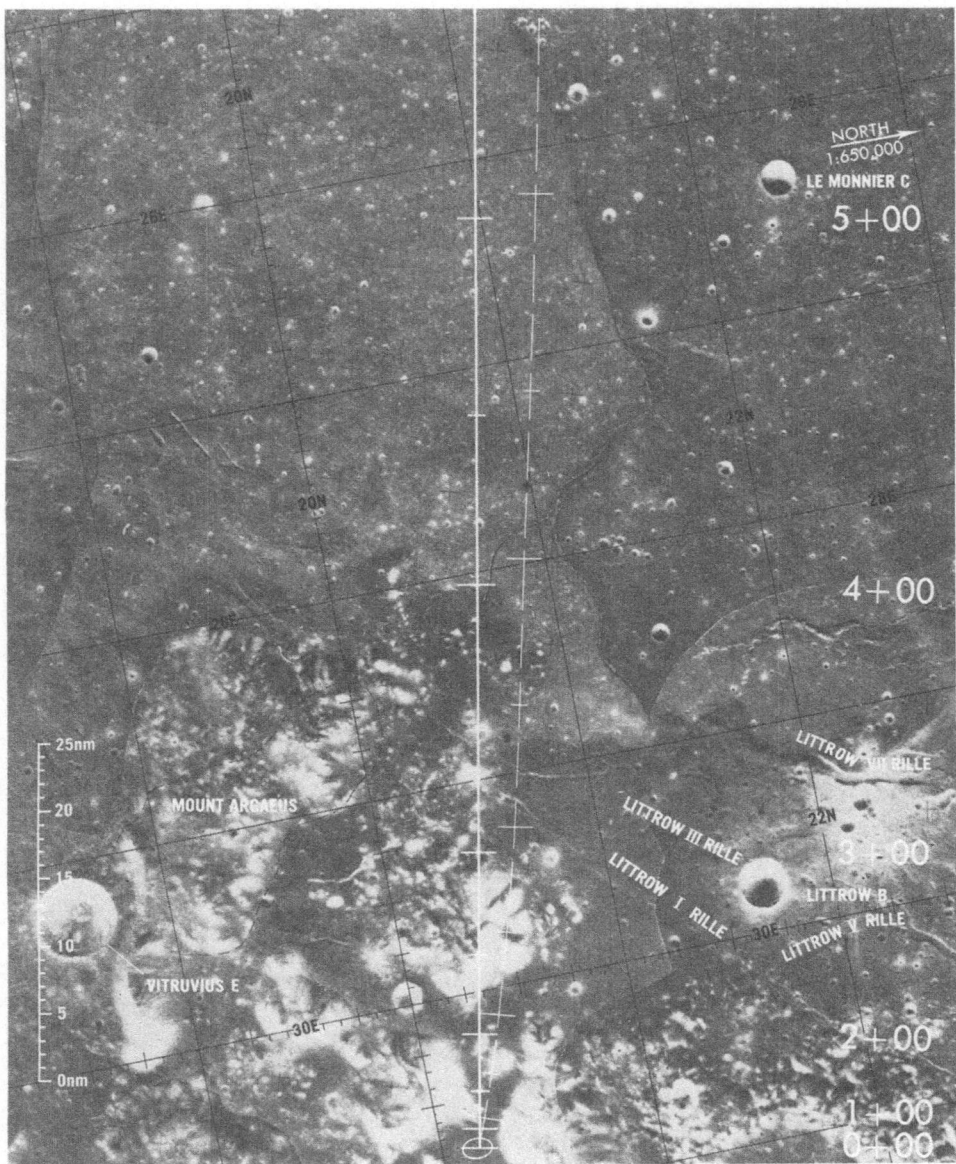

Fig. 8.13.   This is panel 1 of the 2-panel Apollo 17 Ascent Monitoring Chart. It is used for monitoring the azimuth and rate of ascent from the landing area.

of the mission, the astronauts used the Lunar Orbital Science Flight Chart (LSF) which was compiled at a scale of 1:2750000 on a Mercator projection. It takes three charts to provide 360° of longitudinal lunar coverage. These are cut into strips 13 in. wide, centered on the orbital path, then taped end to end and folded in the same manner as the CDM chart. Four book-type LSF charts were required for Apollo 17 to

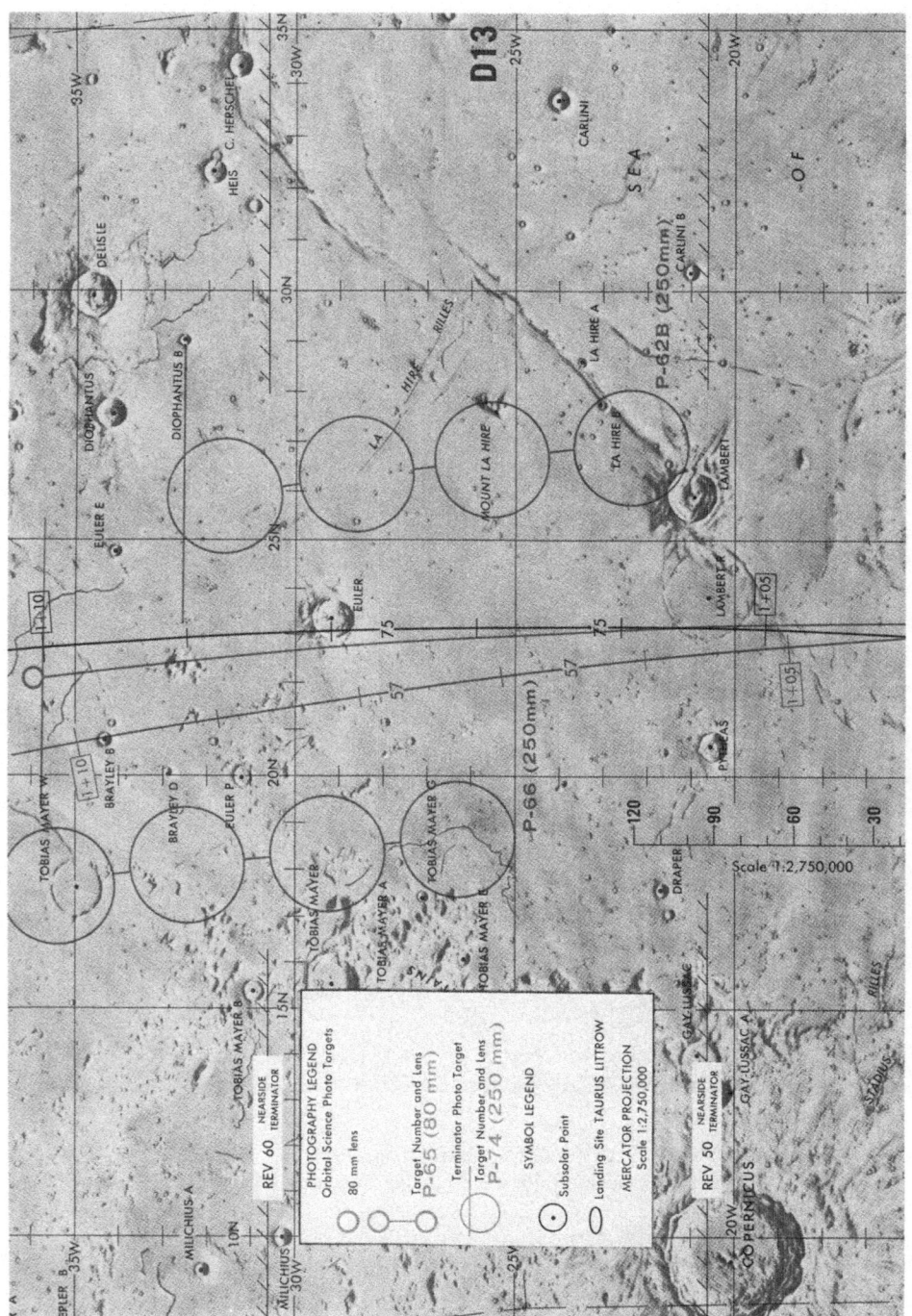

Fig. 8.14.   A section of the Apollo 17 Lunar Orbital Science Flight Chart. Annotated on the chart are some of the preselected areas that were to be photographed with the Hasselblad camera while in lunar orbit.

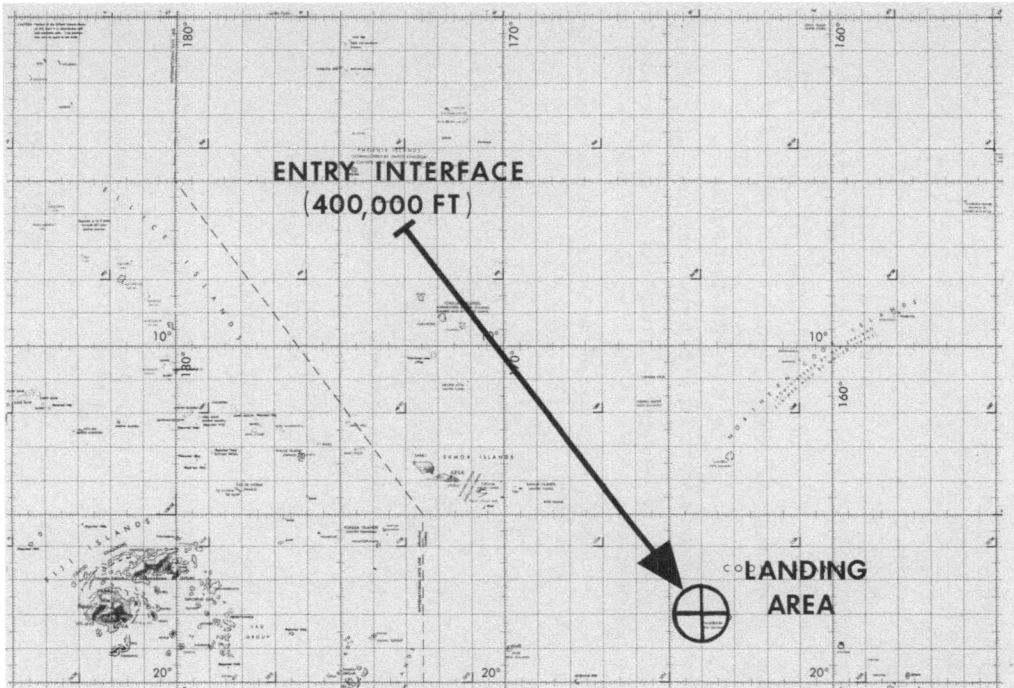

Fig. 8.15. A section of the Pacific Recovery Planning Chart showing the Apollo 17 recovery area. Splashdown on December 19. 1972 terminated the last of the Apollo missions.

display all necessary photographic requirements along the 75 lunar orbits (about 18 orbits per chart).

Figure 8.14 shows a section of one LSF chart which displays some of the targets that were to be photographed during the last 18 revolutions.

3.6. RECOVERY

Splashdown of Apollo 17 in the South Pacific on December 19, 1972 terminated the last of the Apollo missions. To support Earth recovery operations which started back with Mercury, ACIC produced a 1:10000000 scale Pacific Recovery Planning Chart (NPP) on a Mercator projection. A section of this chart is shown in Figure 8.15, annotated with the azimuth re-entry ground track commencing at an altitude of 400000 ft and terminating at 18° S and 166° W in the Cook Islands, approx 2400 nm south of Hawaii.

To cover the possibility of an emergency landing during the Earth orbit phase, both NASA and DOD Recovery Forces were furnished other Recovery Charts produced for Mercury and Gemini that provide coverage of the equatorial band between 40° N–S latitude.

Reference

Carder, Robert W.: 1972, 'Unique Charts for Space Missions', NAVIGATION: *Journal of The Institute of Navigation* **19**, No. 4.

HIGHLIGHTS OF MANNED SPACE FLIGHTS
(NASA Publication EP-91)

	Date	Flight time (hrs:min:s)	Revo-lutions	Spacecraft name	Remarks
Mercury					
Alan B. Shephard, Jr.	5/5/61	00:15:22	Suborbital	Freedom 7	America's first manned space flight.
Virgil I. Grissom	7/21/61	00:15:37	Suborbital	Liberty Bell 7	Evaluated spacecraft functions.
John H. Glenn, Jr.	2/20/62	04:55:23	3	Friendship 7	America's first manned orbital space flight.
M. Scott Carpenter	5/24/62	04:56:05	3	Aurora 7	Initiated research experiments to further future space efforts.
Walter M. Schirra, Jr.	10/3/62	09:13:11	6	Sigma 7	Developed techniques and procedures applicable to extended time in space.
L. Gordon Cooper. Jr.	5/15–16/63	34:19:49	22	Faith 7	Met the final objective of the Mercury program - spending one day in space.
Gemini					
Virgil I. Grissom John W. Young	3/23/65	04:52:31	3	Gemini 3	America's first two-man space flight.
James A. McDivitt Edward H. White, II	6/3–7/65	97:56:12	62	Gemini 4	First 'walk in space' by an American astronaut. First extensive maneuver of spacecraft by pilot.
L. Gordon Cooper, Jr. Charles Conrad, Jr.	8/12–29/65	190:55:14	120	Gemini 5	Eight day flight proved man's capacity for sustained functioning in space environment.
Frank Borman James A. Lovell, Jr.	12/4–18/65	330:35:01	206	Gemini 7	World's longest manned orbital flight.
Walter M. Schirra, Jr. Thomas P. Stafford	12/15–16/65	25:51:24	16	Gemini 6A	World's first successful space rendezvous.
Neil A. Armstrong David R. Scott	3/16–17/66	10:41:26	6.5	Gemini 8	First docking of two vehicles in space.
Thomas P. Stafford Eugene A. Cernan	6/3–6/66	72:20:50	45	Gemini 9A	Three rendezvous of a spacecraft and a target vehicle. Extravehicular exercise – 2 h 7 min.

HIGHLIGHTS OF MANNED SPACE FLIGHTS (*continued*)

	Date	Flight time (hrs:min:s)	Revo-lutions	Spacecraft name	Remarks
John W. Young Michael Collins	7/18–21/66	70:46:39	43	Gemini 10	First use of target vehicle as source of propellant power after docking. New altitude record – 475 miles.
Charles Conrad, Jr. Richard F. Gordon, Jr.	9/12–15/66	71:17:08	44	Gemini 11	First rendezvous and docking in initial orbit. First multiple docking in space. First formation flight of two space vehicles joined by a tether. Highest manned orbit – apogee about 853 miles.
James A. Lovell, Jr. Edwin E. Aldrin. Jr.	11/11–15/66	94:34:31	59	Gemini 12	Astronaut walked and worked outside of orbiting spacecraft for more than $5\frac{1}{2}$ hr – a record proving that a properly equipped and prepared man can function effectively outside of his space vehicle. First photograph of a solar eclipse from space.
Apollo Walter H. Schirra Donn Eisele Walter Cunningham	10/11–22/68	260:8:45	163	Apollo 7	First manned Apollo flight demonstrated the spacecraft, crew and support elements. All performed as required.
Frank Borman James A. Lovell, Jr. William Anders	12/21–27/68	147:00:41	10 rev. of Moon	Apollo 8	History's first manned flight to the vicinity of another celestial body.
James A. McDivitt David R. Scott Russell L. Schweickart	3/3–13/69	241:00:53	151	Apollo 9	First all-up manned Apollo flight (with Saturn V and command. service, and lunar modules). First Apollo EVA. First docking of of CSM with LM.
Thomas P. Stafford John W. Young Eugene A. Cernan	5/18–26/69	192:03:23	31 rev. of Moon	Apollo 10	Apollo LM descended to within 9 miles of Moon and later rejoined CSM. First rehearsal in lunar environment.
Neil A. Armstrong Michael Collins Edwin E. Aldrin, Jr.	7/16–24/69	195:18:35	30 rev. of Moon	Apollo 11	First landing of men on the Moon. Total stay time: 21 hr, 36 min.
Charles Conrad, Jr. Richard F. Gordon, Jr. Alan L. Bean	11/14–24/69	244:36:25	45 rev. of Moon	Apollo 12	Second manned exploration of the Moon. Total stay time: 31 hr 31 min.

HIGHLIGHTS OF MANNED SPACE FLIGHTS *(continued)*

	Date	Flight time (hrs:min:s)	Revo-lutions	Spacecraft name	Remarks
James A. Lovell, Jr. John L. Swigert, Jr. Fred W. Haise, Jr.	4/11–17/70	142:54:41	–	Apollo 13	Mission aborted because of service module oxygen tank failure.
Alan B. Shepard, Jr. Stuart A. Roosa Edgar D. Mitchell	1/13–2/9/71	216:01:59	34 rev. of Moon	Apollo 14	First manned landing in and exploration of lunar highlands. Total stay time: 33 hr 31 min.
David R. Scott Alfred M. Worden James B. Irwin	7/26–8/7/71	295:12:00	74 rev. of Moon	Apollo 15	First manned landing to carry an electric-powered lunar roving vehicle. Total EVA travel distance: 7.8 miles. Total stay time: 66 hr 56 min.
John W. Young Thomas K. Mattingly Charles M. Duke, Jr.	4/16–4/27/72	265:51:00	65 rev. of Moon	Apollo 16	Fifth manned lunar landing and first trip to the southern highlands. Total stay time: 71 hr 02 min.
Eugene A. Cernan Ronald E. Evans Harrison H. Schmitt	12/6–12/19/72	301:52:00	75 rev. of Moon	Apollo 17	Final Apollo mission and first one to carry a geologist for a first-hand study of the Moon. Total stay time: 74 hr 59 min.

CHAPTER 9

U.S. ARMY LUNAR MAPPING

In November 1958, the U.S. Army Map Service (AMS)* conducted a feasibility study to determine production requirements for topographic maps of the visible side of the Moon.

Based on this study AMS decided to produce a two-sheet 1:5000000 scale lunar map compiled on a modified stereographic projection with 1000-m contours and 500-m supplementary contours. To accomplish this project, use was to be made of the best available earth-based telescopic lunar photographs in conjunction with stereophotogrammetric plotting equipment available at the AMS.

The decision to compile a contoured map of the entire visible surface of the Moon by stereophotogrammetric means was a novel idea but it brought AMS many perplexing technical problems which had to be solved. Many scientific and technical disciplines were involved in the course of this efforts; however, one of the main contributors was AMS staff photogrammetrist Albert Nowicki who headed the AMS task force to contour the Moon.

1. Stereo Photographs

The procurement of ideal stereoscopic photographs of the Moon from earth-based telescopes is not possible. For example, simultaneous coverage from two telescopes on opposite sides of the Earth would result in an angular coverage of but 1°44′. This is much too small for precise stereoscopic measurements. However, AMS decided to take advantage of the Moon's libration in longitude of ±7°44′ by choosing stereo pairs of widely librated exposures. They visited various observatories in the United States and Europe and reviewed several thousand photographs. Eventually eight pairs of photographs were selected from the Paris Observatory with dates from March 1896 to January 1907. Unfortunately, emulsions used in those days were very coarse-grained; hence excessive enlargements were limited.

Precise libration characteristics were determined for each of the eight pairs of photographs used for the stereo compilation. Electronic computations were used to establish coordinates for a selenographic grid or 'bird-cage' at 10° intervals, as shown in Figure 9.1. This bird-cage grid was superimposed on one of each pair of lunar images to be projected in the compiling instrument. The selenographic pro-

* The U.S. Army Map Service (AMS) became the U.S. Army Topographic Command (TOPOCOM) in September 1968. Subsequently, in July 1972, TOPOCOM was changed to the Defense Mapping Agency Topographic Center (DMATC).

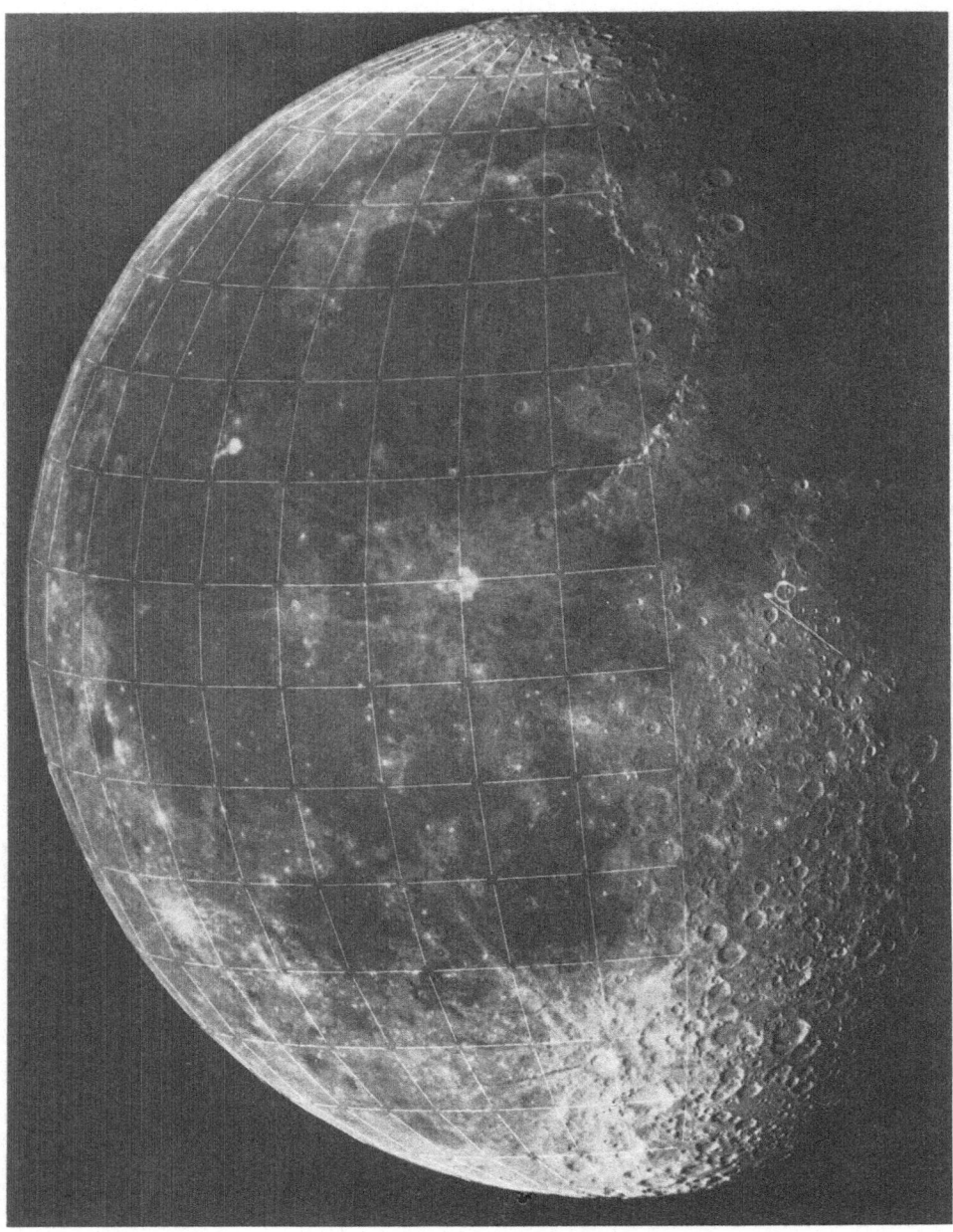

Fig. 9.1. One of the Paris Observatory lunar plates used by AMS to stereo compile a 1000-m contour map of the Moon. Superimposed is a $10° \times 10°$ selenographic grid, nicknamed 'bird-cage' by AMS.

jection made it possible to place all known points on the same horizontal datum. A best overall fit was made to the Austrian astronomer, Schrutka-Rechtenstamm control, and this solution was used to extend the control to areas which were devoid of control points.

2. Lunar Stereoscopy

Figure 9.2 shows the high precision AMS, M-2 stereophotogrammetric plotter designed for use with standard 6-in. focal length, 9 × 9-in. format aerial photographs. Its base height ratio is 0.65 corresponding to an angular coverage of about 36°. These design factors were considerably different from the very long focal length of telescopic lunar photos taken through an unknown lens with undefined distortion,

Fig. 9.2.   M-2 stereophotogrammetric plotter designed for use with 6-in. focal length, 9″ × 9″ format size aerial photographs. The standard projection distance is 2½ ft.   *AMS Photo*

and with a base height ratio of 0.27. Therefore, the M-2 stereoplotter was modified as shown in Figure 9.3. The projectors were raised from their normal height of 2½ ft to 10 ft above the table; the optical system was changed; illumination was increased; and a transistor-powered cooling system was developed for the increased wattage.

The normal orthographic projection and viewing system could not be used. Instead, a new type of tracing table and platen were fabricated to provide a curved mapping surface which could be adjusted precisely in a direction normal to the surface of the projected spatial lunar model. The platen consists of a precisely ground (ground to correspond to the degree of curvature of the projected space image) lucite surface plate attached to a variable height gauge. Three coatings of a special type of lacquer were applied: the first coat was of white lacquer to reflect light within; the second coat was of black lacquer to eliminate the formation of a possible 'hot' spot in the center; the third coat was of white lacquer to provide a reflective viewing surface. Minute holes, at $\frac{1}{4}$-in. intervals, were drilled through the three layers of lacquer, and

Fig. 9.3.   M-2 stereophotogrammetric plotter modified by AMS to accomodate lunar photographs. The projection distance was increased to 10 ft.   *AMS Photo*

with light piped to the lucite platen, a floating reference plane was established. Variations in elevations above or below an assumed datum of the projected spatial model were recorded on a calibrated micrometer scale.

3. Compilation Procedure

In the compilation stage, a sheet of frosted stable-base material was fastened to the curved surface of the lucite platen. The platen, as well as the sheet, was large enough to include all of the area projected into space of any of the 288 $10° \times 10°$ segments which made up the surface of the stereoscopic image of the Moon. Corner ticks and the graticule lines for specific $10° \times 10°$ squares were drawn on the frosted sheet. Edge ties from previously compiled squares were also added. After the platen was properly oriented to the control and edge tie material, and the vertical vernier scale of the platen was indexed to a specific contour interval, compilation at a scale of 1:3 300 000 was accomplished as shown in Figure 9.4. Much of the stereo compilation on this Army project was accomplished by AMS photogrammetrist Felix Bizzoco.

Since the original compilation was a spherical projection and the map was to be published on a modified stereographic projection, it was necessary to rectify each

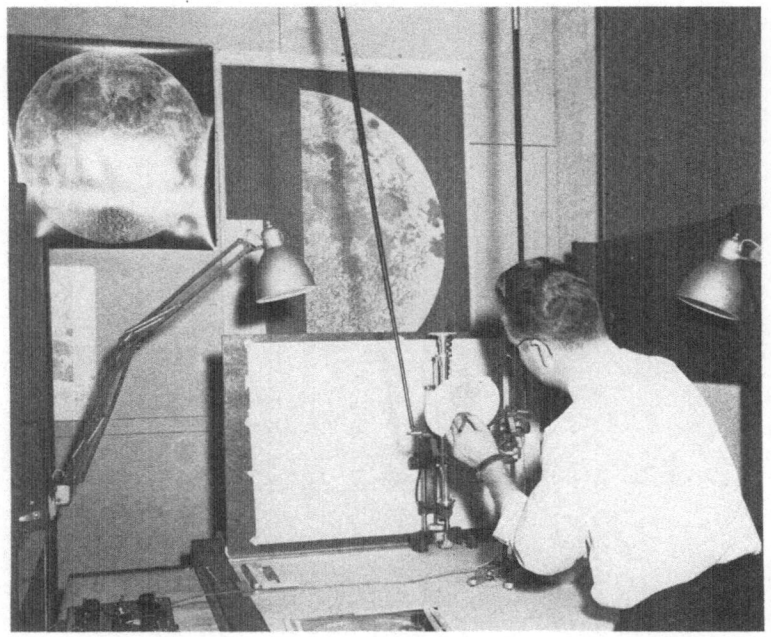

Fig. 9.4.   AMS stereo operator using the modified M-2 stereophotogrammetric plotter to compile individual 10° × 10° segments which were then rectified and joined to form the 1:3 300 000 scale compilation of the entire lunar Earth side.   *AMS Photo*

10° square. After each square had been rectified and additional detail added from large-scale photos, the map was ready for drafting or color separation to put it into final form for reproduction.

4. Topographic Maps

From the lunar stereo compilation AMS produced three series of small-scale lunar maps. The first map to be published, shown in Figure 9.5 was a 1:2 500 000 scale sheet covering the Mare Nectaris-Mare Imbrium area. This sheet, centered on 40° N–S lat., 40° E–W long., 32″ × 54″, was issued in March 1962 in a gradient tint and a shaded relief style.

In September 1964, the entire near side was published in two sheets at the scale of 1:5 000 000. Each sheet measured 38″ × 54″ and was published in three versions: a relief, a gradient tint and a shadient relief rendition. The two sheets (eastern and western sections) are showing in Figure 9.6. The area surrounding the crater Coper-

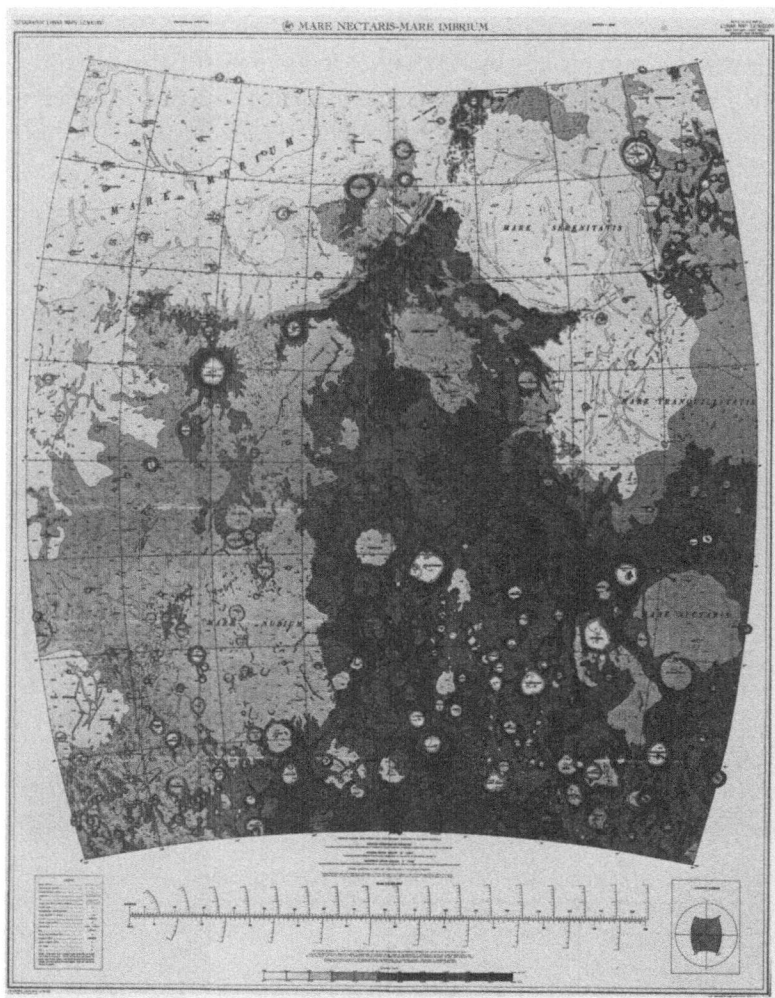

Fig. 9.5.   AMS 1:2 500 000 scale lunar map covering the Mare Nectaris–Mare Imbrium region. It was published in March 1962.

nicus is shown in Figure 9.7 at true scale. Note that small craters are symbolized by a small circle with a dot in the center.

The vertical datum is based on Aristarchus as being the lowest of all craters listed in the Schrutka points. Therefore, a value of zero was chosen for the crater Aristarchus to avoid the use of negative contour values. This is equivalent to a radius value at this point of 1732.4 km; for Mösting A the radius is 1739.4 km, corresponding to an elevation expressed as a contour value of 7000 m. This value for Mösting A was chosen as the fundamental vertical reference to which all other elevations were referenced. Contours are shown at a 1000-m interval with 500-m supplementary contours in the maria.

In 1965, the above three versions were published at a scale of 1:2000000 in a six-sheet series, each sheet measuring 38" × 54". Shown in Figure 9.8 is Sheet 1 of this series covering the northwestern area of the lunar near side.

Also a 1:5000000 scale photomosaic of the entire near side, on an orthographic projection was lithographed in color, AMS stated that this map served a two-fold

Fig. 9.6.   This is a composite of the AMS, two sheet, 1:5000000 scale Lunar Topographic Map. It was published in September 1964.

purpose: to make available a pictorial reference to the entire visible surface on one sheet; and to serve as a guide to the pictorial artists in rendering the lunar surface features on the two-sheet version of the lunar topographic map. A copy of this map is shown in Figure 9.9. It is made from two lunar photographs, a first and last quarter

Moon, joined in the center, and then tinted in green, blue and purple for the maria and yellow and brown for the highlands.

Shown in Figure 9.10 is one of four 1:250000 scale topographic maps produced in the Landsberg area. These maps were published in March 1964 based on an enlargement of the 1961 1:3300000 stereophotogrammetric compilation supplemented by visual interpretation from Lick and Yerkes Observatory photographs taken in

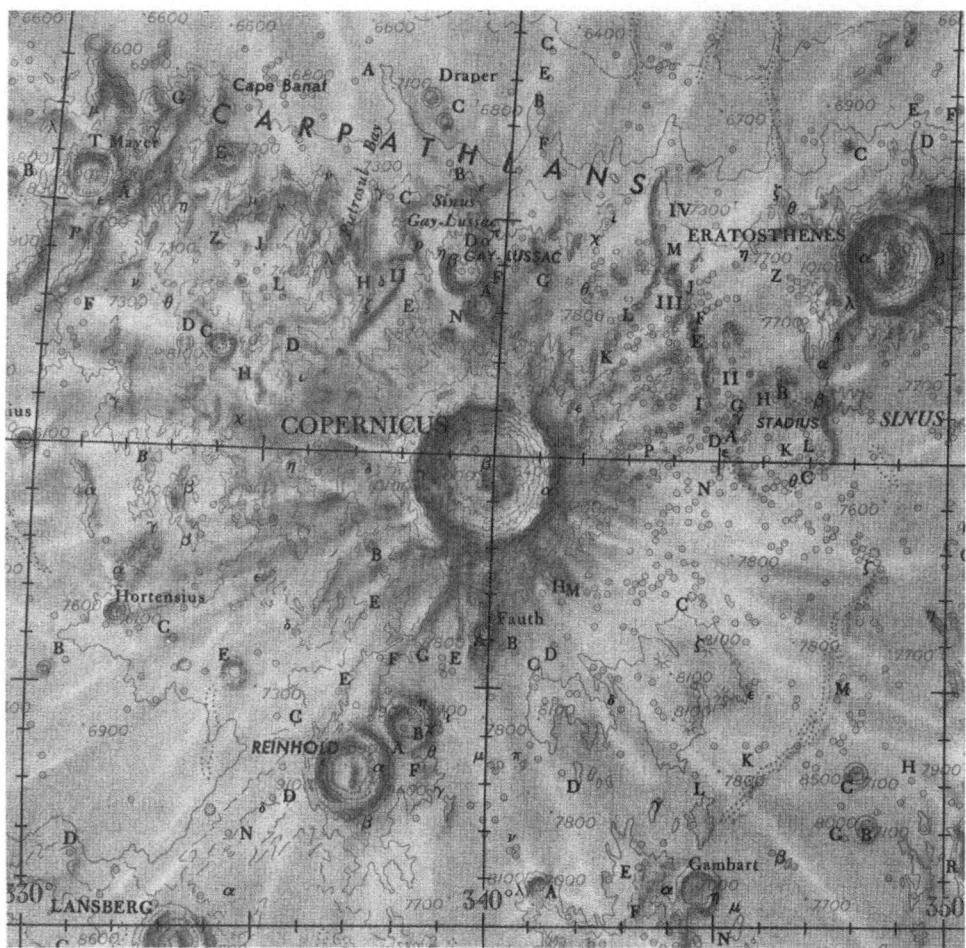

Fig. 9.7.   A true scale reproduction of the Copernicus area from the AMS Lunar Topographic Map
shown in Figure 9.6.

1936. Contours are shown at a 250-m interval. This particular area was selected for mapping because it seemed to NASA, at that time, to be a good candidate area for a manned landing. However, mapping at this scale from telescopic photographs proved to be beyond the resolving power of the telescope and no additional lunar maps using this technique and source were attempted.

5. Ranger Mapping

In 1966, AMS initiated a research project under the direction of AMS photogram-metrist Donald Light to compile a stereo topographic map from Ranger VIII photographs. Ranger VIII returned 7137 photos in February 1965 and provided a means for experimenting with photographic records taken by a television vidicon system and subsequently transmitted to Earth. Because the descending trajectory caused

Fig. 9.8.   Sheet 1 of the six sheet, 1:2 000 000 scale, AMS Lunar Topographic Map published in 1965. This sheet covers the northwestern area of the near side of the Moon.

the photographs to be taken in a telescoping manner as the spacecraft approached the Moon, very little base distance between pictures existed and, therefore, photo-grammetric methods were seriously hampered.

A Ranger stereo model could not be absolutely oriented (leveled) in the plotter; therefore, AMS resorted to a computational method to compile the contours. A dense network of x, y, z stereoplotter coordinates were measured from the Ranger stereo models and recorded in digital form. These coordinates were then adjusted by ana-lytical methods to the selenodetic control to obtain a dense network of terrain data points. Subsequently a computer contouring technique was employed to interpolate contours.

Two maps of the Sabine crater area were compiled at 1:250 000 with 100-m

Fig. 9.9.   A 1:5000000 scale Pictorial Map of the lunar near side published by AMS in November 1963.

contours. One of these maps, $13'' \times 20''$, is shown in Figure 9.11. Also a larger scale, 1:50000 map was made in the same area and titled 'Mare Tranquillitatis'.

6. Orbiter Mapping

From Orbiter Missions I, II and III, described in the chapter on U.S. Air Force Lunar Mapping, AMS produced 1:100000 scale Topographic Maps over twelve Orbiter sites which are listed at the end of this chapter. They also published companion 1:100000 scale Photomaps of the same areas. In addition, 1:25000 scale Photomaps were compiled for six of the twelve sites. This mapping was accomplished from April 1967 through June 1968.

The Lunar Photomap series reflects only photomosaic detail and is lithographed in black. The Lunar Topographic Map series shows hypsography by shadient relief,

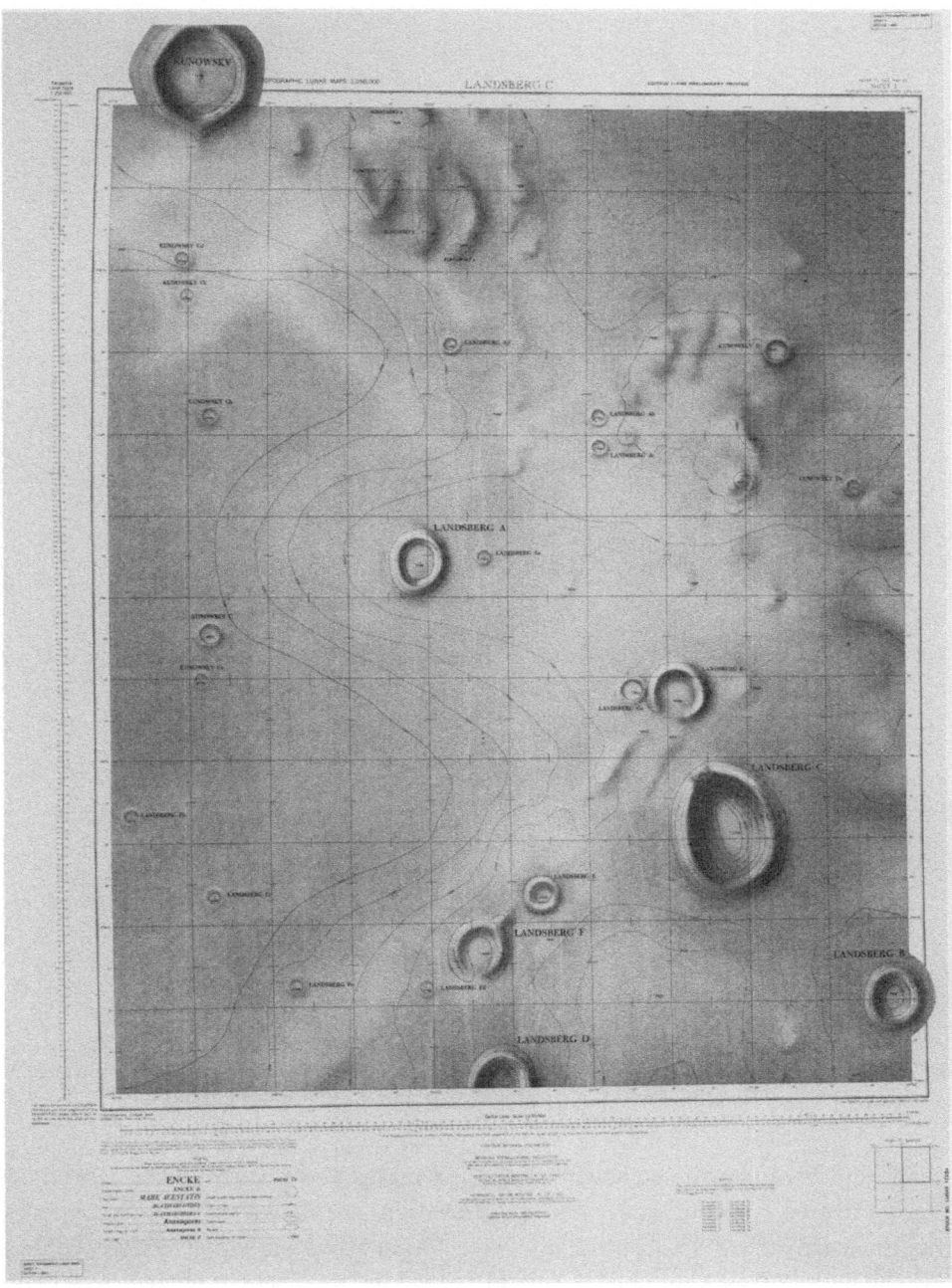

Fig. 9.10. One of four 1:250000 scale lunar topographic maps of the Landsberg area published by AMS in March 1964.

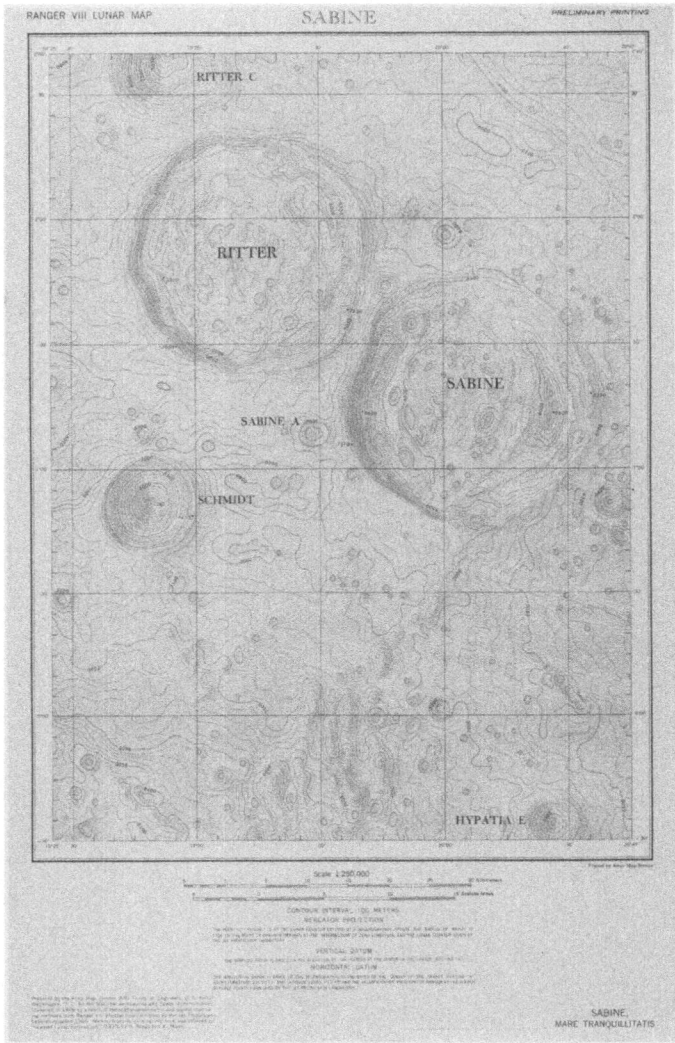

Fig. 9.11.   A 1:250000 scale AMS lunar topographic map of the Sabine area compiled by stereo methods
from Ranger VIII photographs. Contours are at a 100-m interval.

contours, spot elevations, height or rim elevations above surrounding terrain, and
crater depths. Contour values and spot elevations are shown as lunar radius vectors
expressed in meters. The contour interval is 100-m with 50-m supplementary contours
in the flatter areas. Contour information is lacking in areas not having stereoscopic
coverage. Reliability information is expressed in the map margin. The map base is a
green and black duotone printing. Relief values are overprinted in black with contours in red.

AMS employed stereoplotting equipment to compile feature positions and contour
information for the Topographic Map manuscripts. However, the segmented framelet

form of Orbiter photographs prevented continuous compilation within each stereo model, requiring overall adjustment of individual framelet model compilations.

Shown in Figure 9.12 is the Topographic Map ORB-I-1 (100), size 21″ × 41″, which was published in September 1967. This site map, compiled on a Mercator projection, is located just south of the equator from 40°00′ to 43°48′ E long. A note in the border

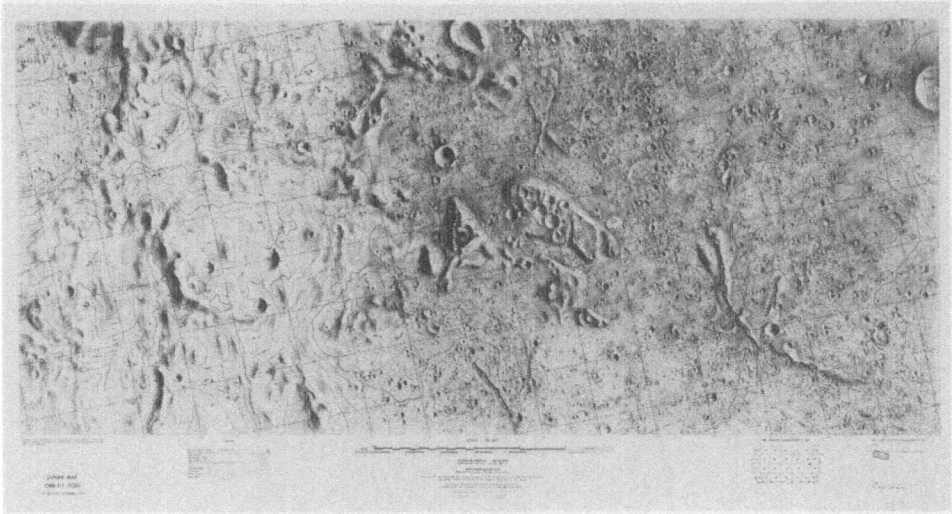

Fig. 9.12.   Lunar Map ORB-I-1 (100) – one of the first 1:100000 scale lunar topographic maps produced by AMS from Orbiter photographs. It was published in September 1967.

states that the vertical and horizontal control was established by photogrammetric triangulation using orbit constraints.

7. Surveyor Site Mapping

Surveyor III was launched on April 17, 1967 and landed on April 19th in Oceanus Procellarum 3°12′04″ S lat., 23°22′54″ W long. When the spacecraft landed, it came to rest on the inside of a crater giving it a 12.4° tilt from the local vertical. This crater became an object of intense interest to the scientific community and NASA requested AMS to prepare two maps of the crater and surrounding areas.

Two Lunar Orbiter III high-resolution convergent photographs, H137 and H154, were used to compile a 1:2000 scale Photomap and a 1:500 scale Topographic Map. The two high-resolution photographs were controlled to a triangulation previously derived from Lunar Orbiter I medium-resolution photographs covering the same area. The map area was triangulated by a block process to establish a large number of pass points which in turn controlled the AS-11A analytical plotter upon which the 1:2000 scale manuscript with 2-m contours was compiled. AMS personnel responsible for this work were Charles Shull and Lynn Schenk.

The manuscript was used to control the 2000 scale Photomap which was published with 10-m contours in January 1968. For the larger scale Topographic Map, the original manuscript was enlarged to 1:500 and additional features were added from a large scale blow-up of the high-resolution photo coverage.

Shown in Figure 9.13 is the 1:500 scale Surveyor III Site Map containing 2-m contours. This map was published in February 1968.

Previously, in October 1967, AMS published a 1:100 scale Pictorial Map of the

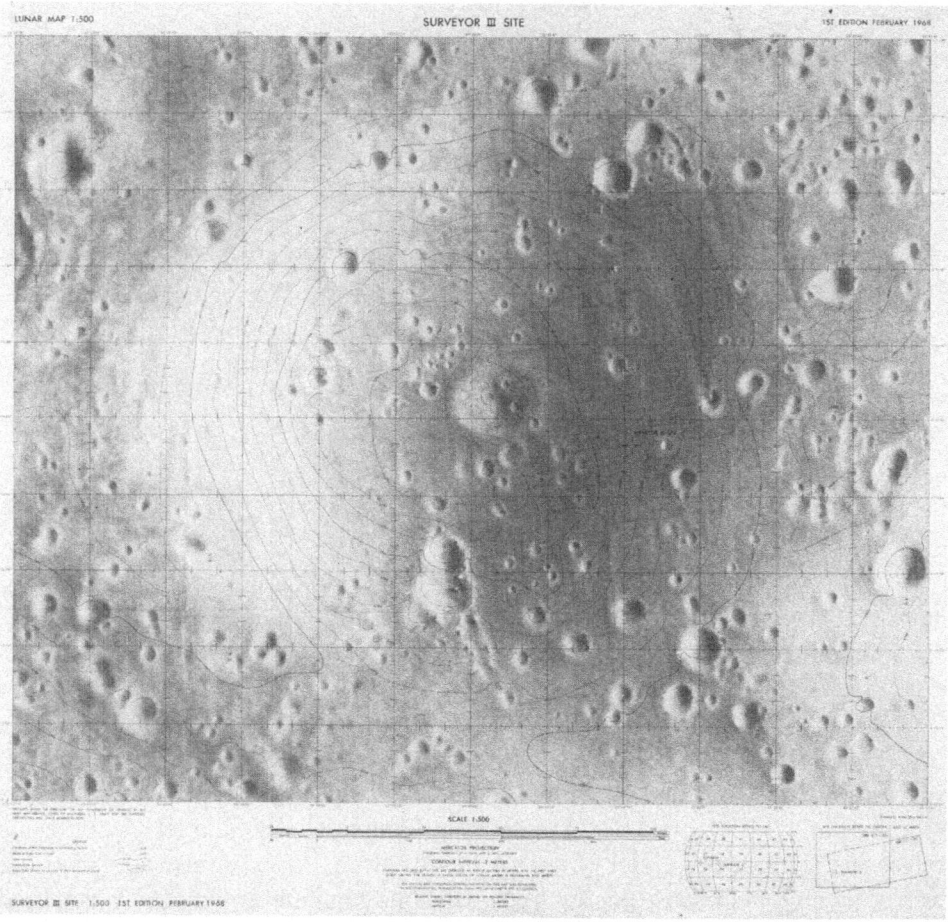

Fig. 9.13.   A 1:500 scale lunar map of the Surveyor III landing area compiled from Orbiter III photographs. It was published by AMS in February 1968.

Surveyor I Site. This map is compiled on an orthographic projection based on the local lunar surface and the Surveyor I Site as its center. The extent of coverage is 46 m in any direction from the Surveyor I camera station. This map does not contain contour or relative height information. In March 1969, TOPOCOM published a 1:1000 scale Photomap of the Surveyor VI landing area based on a compilation made at the NASA Manned Spacecraft Center from photometric reduction of Orbiter video tapes.

8. Scientific Site Mapping

From Lunar Orbiter V photographs TOPOCOM published 1:250000 scale maps covering ten areas of scientific interest. For each site, three types of lunar maps were produced, a Photomap (without contours), a Topographic Photomap (with contours) and a Topographic Map. Publication of individual sheets began at TOPOCOM in June 1969 and was completed by January 1972.

The Lunar Photomap Series shows photomosaic detail and is lithographed in black. The Lunar Topographic Photomap series shows photomosaic detail, over-printed by contours in red; spot elevations are expressed as lunar radius vectors in meters. The Lunar Topographic Map series expresses relief by shadient portrayal, contour and spot elevations. The map base is printed in a duotone green and black. Contours at a 400-m interval with 200-m supplementary contours are printed in red. Contours are not shown in the photo imagery areas lacking stereoscopic coverage.

Shown in Figure 9.14 is the Lunar Topographic Map Gassendi, Sheet A, compiled

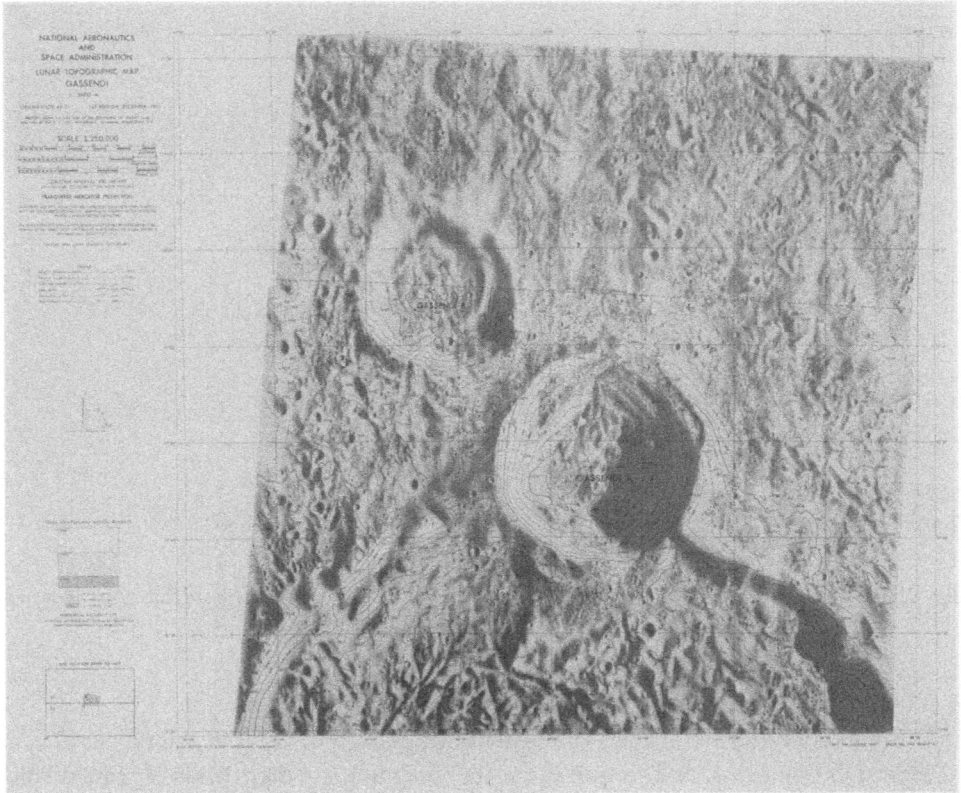

Fig. 9.14. Lunar Topographic Map, Gassendi, Sheet A, compiled from Orbiter V photographs by AMS in December 1971.

in December 1971, from Orbiter V-Site 43.2 photographs. Sheet A covers the northern
portion of the crater Gassendi; Sheet B, the southern section. In Figure 9.14 the
smaller crater Gassendi A is in the center with Gassendi B to the upper left. Running
through Gassendi B is a line which represents the limits of stereo coverage which

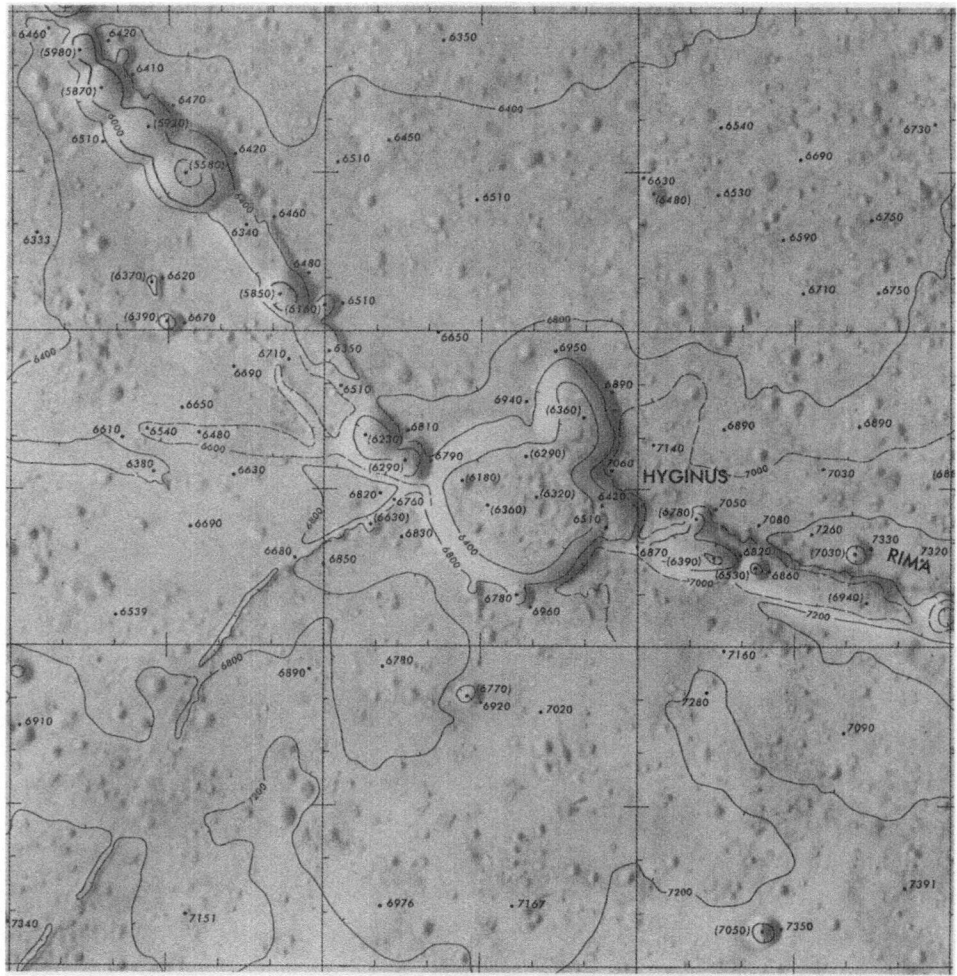

Fig. 9.15. A section of the TOPOCOM produced 1:250000 scale Lunar Topographic Map showing
part of the Hyginus Rille.

means that contours are not shown north of this point. The contour interval is 400 m.
    Shown in Figure 9.15 is a section of another scientific site map, Rima Hyginus. This
Lunar Topographic Map, scale 1:250000, further identified as Orbiter V-Site 23.1
was compiled in November 1970. Shown in Figure 9.16 is a section of the Lunar

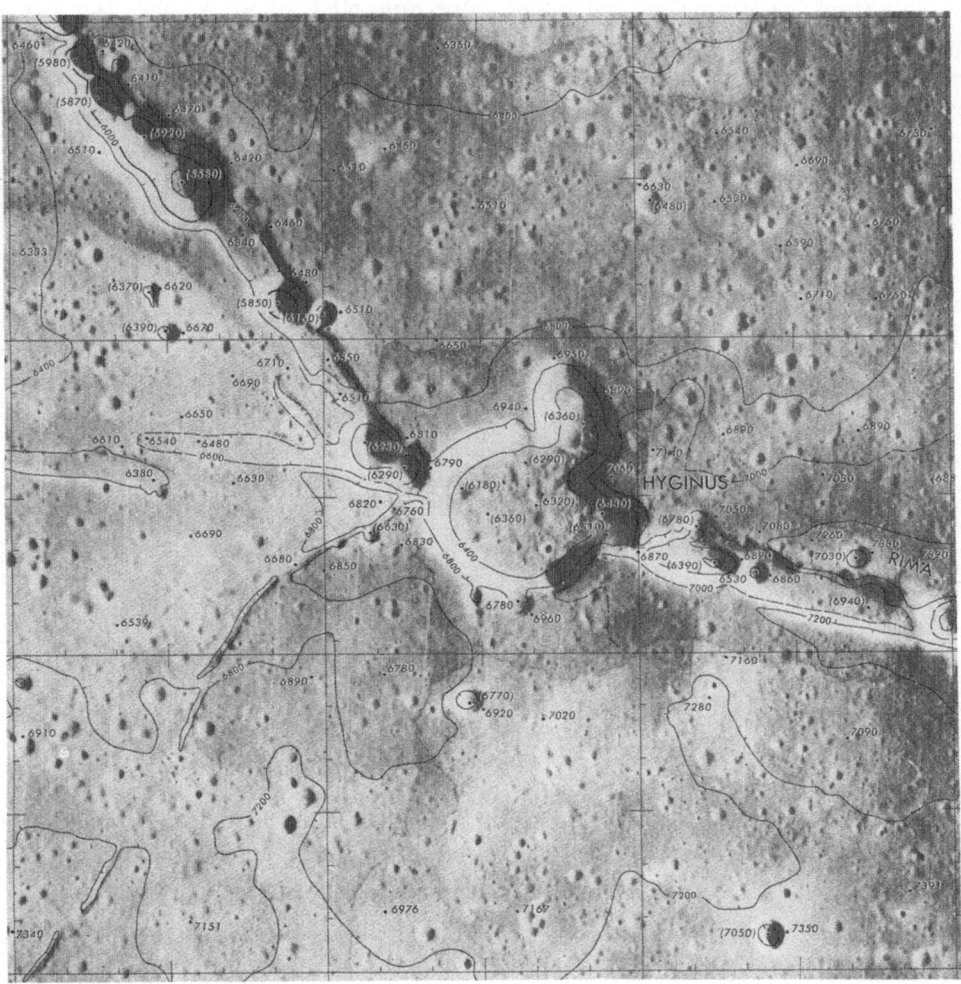

Fig. 9.16. A section of the TOPOCOM produced 1:250000 scale Lunar Topographic Photomap showing part of the Hyginus Rille (same area as Figure 9.15).

Topographic Photomap covering the same area of Rima Hyginus. The contours and spot elevations are identical to the Topographic Map.

9. Lunar Equatorial Zone Mosaic

In November 1969, TOPOCOM published a four-sheet 1:5000000 scale mosaic of the entire 360° lunar equatorial zone from 20° N to 20° S lat. This mosaic was prepared from both medium- and high-resolution prints of the Lunar Orbiters I to V photographs. The four sheets in this series cover the Earth Side (Sheet 1), Eastern Limb (Sheet 2), Far Side (Sheet 3) and the Western Limb (Sheet 4). The projection and major feature names are printed in white by a hold-out masking technique. Shown in Figure

9.17 in Sheet 1, covering the Earth Side Zone from 50° E to 50° W long. The sheet measures 24″ × 45″.

Fig. 9.17.   Lunar Equatorial Zone Mosaic – one of four sheets covering the equatorial region from 20° N to 20° S latitude and published by TOPOCOM in November 1969.

10. Apollo Site Mapping

On Apollo Missions 8, 10, 11, 12 and 14, on board modified Hasselblad 500 EL cameras photographed potential landing sites and other areas of scientific interest. In some areas these photographs replaced existing Orbiter coverage for preflight training and on board flight charts. During the flight of Apollo 14 the astronauts photographed the Descartes region which eventually became the Apollo 16 landing site.

In support of the Apollo 16 mission, NASA requested TOPOCOM to compile a series of four maps of the Descartes region. These were a 1:100000 scale Photomap published in October 1971, a 1:25000 scale Topographic Photomap published in January 1972, a 1:25000 scale and a 1:100000 scale Topographic Map published in April 1972. The horizontal and vertical control was established by photogrammetric triangulation using orbit constraints and is based on Apollo 14 mission support datum (ephemeris) dated April 1971. All four maps were compiled on a transverse Mercator projection with 10-m and 100-m contours for the 1:25000 and 1:100000 Topographic Map, respectively. Shown in Figure 9.18 is a section of the 1:25000 Topographic Map containing the Apollo 16 landing site.

Apollo Missions 15, 16 and 17 provided high-quality mapping photographs along with precise supporting data for about 20% of the lunar surface. The instruments flown comprised a metric camera system, and a panoramic camera which was described in Chapter 6, U.S. Air Force Lunar Mapping.

Fig. 9.18.   A section of the 1:25000 scale Lunar Topographic Map covering the Descarte region. Arrow points to the location of the Apollo 16 landing site. This map was published by TOPOCOM in April 1972.

The photographs taken on the flight of Apollo 15 were used by TOPOCOM to map the Apollo 17 landing area, named Taurus-Littrow. From metric photographs 968, 970, 972, 974, 1655 and 1657 taken in August 1971, they produced a 1:250000 scale Topographic Orthophotomap on a transverse Mercator projection. (An ortho-photomap is made from one or more orthophotographs which is a photographic copy prepared from a perspective photograph in which the displacement of image due to tilt and relief have been removed. Special equipment such as a AS-11A/Geigas Zeiss instrument is required to produce an orthophoto.)

The horizontal and vertical control used on the Taurus-Littrow 1:250000 scale Topographic Orthophotomap was established by photogrammetric triangulation

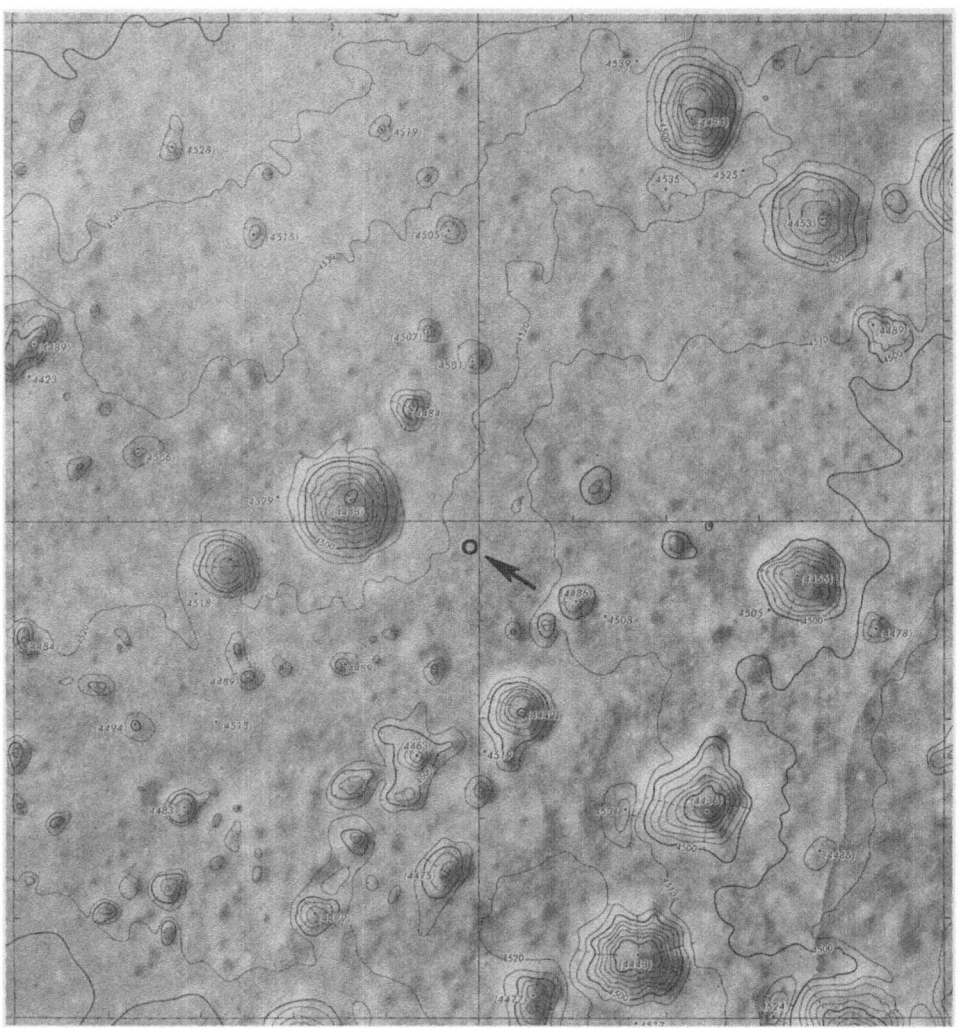

Fig. 9.19.   A section of the 1:25 000 scale Lunar Topographic Photomap covering the Taurus-Littrow region. Arrow points to location of the Apollo 17 landing site. This map was published by TOPOCOM in September 1972.

using orbit constraints and is based on Apollo 15 mission support data (ephemeris) dated December 1971. This gave a horizontal accuracy of ± 36 m (90% probability). This map, printed in halftone screen black shows 100-m contours overprinted in red.

Two additional maps, a 1:50 000 scale and a 1:25 000 scale Topographic Photomap were produced from panoramic frame numbers 9552, 9554, 9557 and 9559. Shown in Figure 9.19 is a section of the 1:25 000 scale Topographic Photomap containing the Apollo 17 landing site. Contours are at a 10-m interval.

References

Light, Donald L.: 1966, 'Ranger Mapping by Analytics', Photogrammetric Engineering: *Journal of the American Society of Photogrammetry*, September.

Nowicki, Albert L.: 1961, 'Topographic Lunar Mapping at the Army Map Service', AMS Technical Report No. 37.

Shull, Charles W. and Schenk, Lynn A.: 1970, 'Mapping the Surveyor III Crater', Photogrammetric Engineering: *Journal of the American Society of Photogrammetry*, June.

LUNAR MAPS PRODUCED BY U.S. ARMY FOR NASA

Description	Sheet No.	Sheet name	Scale	Total sheets	Date	Sheet size
Topographic Mapping						
Lunar Topographic Center Section (Gradient Tint Style)			1:2500000	1	3-62	32" × 54"
Lunar Topographic Center Section (Shaded Relief Style)			1:2500000	1	3-62	32" × 54"
Lunar Pictorial	—		1:5000000	1	11-63	36" × 36"
Lunar Topographic Eastern Section (Relief Style)			1:5000000	1	9-64	38" × 54"
Lunar Topographic Western Section (Relief Style)			1:5000000	1	9-64	38" × 54"
Lunar Topographic Eastern Section (Gradient Tint Style)			1:5000000	1	9-64	38" × 54"
Lunar Topographic Western Section (Gradient Tint Style)			1:5000000	1	9-64	38" × 54"
Lunar Topographic Eastern Section (Shaded Relief Style)			1:5000000	1	9-64	38" × 54"
Lunar Topographic Western Section (Shaded Relief Style)			1:5000000	1	9-64	38" × 54"
Lunar Topographic (Sheet No. 1, 2, 3, 4)			1:250000	4	3-64	30" × 40"
Lunar Topographic (Relief Style)			1:2000000	6	6-65	38" × 54"
Lunar Topographic (Gradient Tint Style)			1:2000000	6	6-65	38" × 54"
Lunar Topographic (Shaded Relief Style)			1:2000000	6	6-65	38" × 54"
Ranger Mapping						
Ranger VIII Lunar Map	—	Sabine	1:250000	1	1966	13" × 20"
Ranger VIII Lunar Map	—	Sabine B	1:250000	1	1966	12" × 15"
Ranger VIII Lunar Map	—	Mare Tranquillitatis	1:50000	1	1966	8" × 11"
Orbiter Site Mapping						
Lunar Photomap	ORB-I-1 (25)	—	1:25000	4	1967	31" × 32"
	ORB-I-1 (100)	—	1:100000	1	6-67	22" × 44"
	ORB-I-2 (100)	—	1:100000	1	2-68	22" × 44"
	ORB-I-3 (100)	—	1:100000	1	4-67	21" × 44"
	ORB-I-4 (100)	—	1:100000	2	2-68	19" × 25"
	ORB-I-7 (100)	—	1:100000	1	6-67	20" × 40"
	ORB-I-8 (100)	—	1:100000	1	6-67	20" × 27"
	ORB-II-2 (25)	—	1:25000	1	11-67	29" × 50"
	ORB-II-2 (100)	—	1:100000	1	9-67	24" × 18"
	ORB-II-6 (25)	—	1:25000	4	11-67	28" × 27"
	ORB-II-6 (100)	—	1:100000	1	10-67	24" × 26"
	ORB-II-11 (25)	—	1:25000	4	11-67	28" × 33"
	ORB-II-11 (100)	—	1:100000	1	11-67	26" × 30"
	ORB-II-12 (25)	—	1:25000	4	11-67	26" × 29"
	ORB-II-12 (100)	—	1:100000	1	11-67	23" × 27"
	ORB-III-9 (25)	—	1:25000	4	2-68	27" × 34"
	ORB-II-13 (25)	—	1:25000	4	11-67	26" × 29"
	ORB-II-13 (100)	—	1:100000	1	11-67	23" × 27"

Description	Sheet No.	Sheet name	Scale	Total sheets	Date	Sheet size
Orbiter Site Mapping						
Lunar Photomap	ORB-III-9 (100)	—	1:100000	1	1-68	23" × 26"
	ORB-III-11 (25)	—	1:25000	2	1-68	29" × 34"
	ORB-III-11 (100)	—	1:100000	1	1-68	23" × 26"
Lunar Map	ORB-I-1 (100)	—	1:100000	1	9-67	21" × 41"
	ORB-I-2 (100)	—	1:100000	1	6-68	21" × 41"
	ORB-I-3 (100)	—	1:100000	1	5-67	21" × 41"
	ORB-I-4 (100)	—	1:100000	2	5-68	19" × 25"
	ORB-1-7 (100)	—	1:100000	1	9-67	20" × 37"
	ORB-1-8 (100)	—	1:100000	1	7-67	20" × 24"
	ORB-II-2 (100)	—	1:100000	1	12-67	21" × 19"
	ORB-II-6 (100)	—	1:100000	1	12-67	23" × 24"
	ORB-II-11 (100)	—	1:100000	1	2-68	26" × 26"
	ORB-II-13 (100)	—	1:100000	1	12-67	24" × 24"
	ORB-III-9 (100)	—	1:100000	1	3-68	24" × 26"
	ORB-III-11 (100)	—	1:100000	1	2-68	21" × 25"
Surveyor Site Mapping						
Surveyor I Site – Map		—	1:100	1	10-67	39" × 41"
Surveyor III Site – Photomap		—	1:2000	1	1-68	24" × 30"
Surveyor III Site – Topographic Map		—	1:500	1	2-68	24" × 28"
Surveyor VI Site – Experimental		—	1:1000	1	3-69	26" × 30"
Scientific Site Mapping						
Lunar Photomap	Orbiter V – Site 12	Censorius	1:25000	2	5-69	28" × 42"
	Orbiter V – Site 12	Censorius	1:250000	1	5-69	18" × 24"
	Orbiter V – Site 14	Rima Littrow	1:250000	1	8-69	20" × 24"
	Orbiter V – Site 23.1	Rima Hyginus	1:250000	1	8-69	18" × 22"
	Orbiter V – Site 24	Hipparchus	1:250000	1	1-70	18" × 20"
	Orbiter V – Site 26.1	Rima Hadley	1:250000	2	4-70	22" × 24"
	Orbiter V – Site 30	Tycho	1:250000	2	10-69	22" × 34"
	Orbiter V – Site 37	Copernicus	1:250000	1	12-69	21" × 24"
	Orbiter V – Site 43.2	Gassendi	1:250000	2	12-71	20" × 26"
	Orbiter V – Site 46	Prinz	1:250000	1	4-70	28" × 32"
	Orbiter V – Site 51	Marius F.	1:250000	1	1-70	22" × 23"

LUNAR MAPS PRODUCED BY U.S. ARMY FOR NASA *(continued)*

Description	Sheet No.	Sheet name	Scale	Total sheets	Date	Sheet size
Scientific Site Mapping						
Lunar Topographic Photomap	Orbiter V – Site 14	Rima Littrow	1:250 000	1	5-70	20″ × 24″
	Orbiter V – Site 23.1	Rima Hyginus	1:250 000	1	10-70	18″ × 22″
	Orbiter V – Site 24	Hipparchus	1:250 000	1	11-70	18″ × 20″
	Orbiter V – Site 26.1	Rima Hadley	1:250 000	2	11-70	22″ × 24″
	Orbiter V – Site 30	Tycho	1:250 000	2	8-71	22″ × 34″
	Orbiter V – Site 37	Copernicus	1:250 000	1	1-71	21″ × 24″
	Orbiter V – Site 43.2	Gassendi	1:250 000	2	10-71	20″ × 26″
	Orbiter V – Site 46	Prinz	1:250 000	1	12-70	28″ × 32″
	Orbiter V – Site 51	Marius F.	1:250 000	1	8-70	22″ × 23″
Lunar Topographic Map	Orbiter V – Site 14	Rima Littrow	1:250 000	1	5-70	20″ × 24″
	Orbiter V – Site 23.1	Rima Hyginus	1:250 000	1	11-70	18″ × 22″
	Orbiter V – Site 24	Hipparchus	1:250 000	1	3-71	18″ × 20″
	Orbiter V – Site 26.1	Rima Hadley	1:250 000	2	1-71	22″ × 24″
	Orbiter V – Site 30	Tycho	1:250 000	2	9-71	22″ × 34″
	Orbiter V – Site 37	Copernicus	1:250 000	1	1-71	21″ × 24″
	Orbiter V – Site 43.2	Gassendi	1:250 000	2	12-71	20″ × 26″
	Orbiter V – Site 46	Prinz	1:250 000	1	5-71	28″ × 32″
	Orbiter V – Site 51	Marius F.	1:250 000	1	4-71	22″ × 23″
Lunar Equatorial Zone Mosaic			1:2 500 000	4	11-69	24″ × 45″
Apollo Site Mapping						
Lunar Photomap		Descartes	1:100 000	1	10-71	25″ × 25″
Lunar Topographic Photomap		Descartes	1:25 000	1	1-72	28″ × 31″
Lunar Topographic Map		Descartes	1:25 000	1	4-72	28″ × 31″
Lunar Topographic Map		Descartes	1:100 000	1	4-72	25″ × 25″
Lunar Topographic Photomap		Taurus – Littrow	1:25 000	1	9-72	43″ × 29″
Lunar Topographic Photomap		Taurus – Littrow	1:50 000	1	9-72	30″ × 32″
Lunar Topographic Orthophotomap		Taurus – Littrow	1:250 000	1	9-72	30″ × 26″

CHAPTER 10

U.S.S.R. LUNAR MAPPING

On October 4, 1959, the third cosmic rocket, Luna 3 was successfully launched in the Union of Soviet Socialist Republics. It orbited around the Moon's far side and photographs taken on 35 mm film with cameras equipped with lenses of 20- and 50- cm focal length were televised back to ground receiving stations after the spacecraft had returned to the vicinity of the Earth.

Thus, the first photographs of the far side of the Moon (see Figure 10.1) were obtained in the Soviet Union. The development of methods for studying the Luna 3 photographs to expose features of the lunar surface, the compilation of a catalog with a description of the characteristics of the new features and the compilation of a far

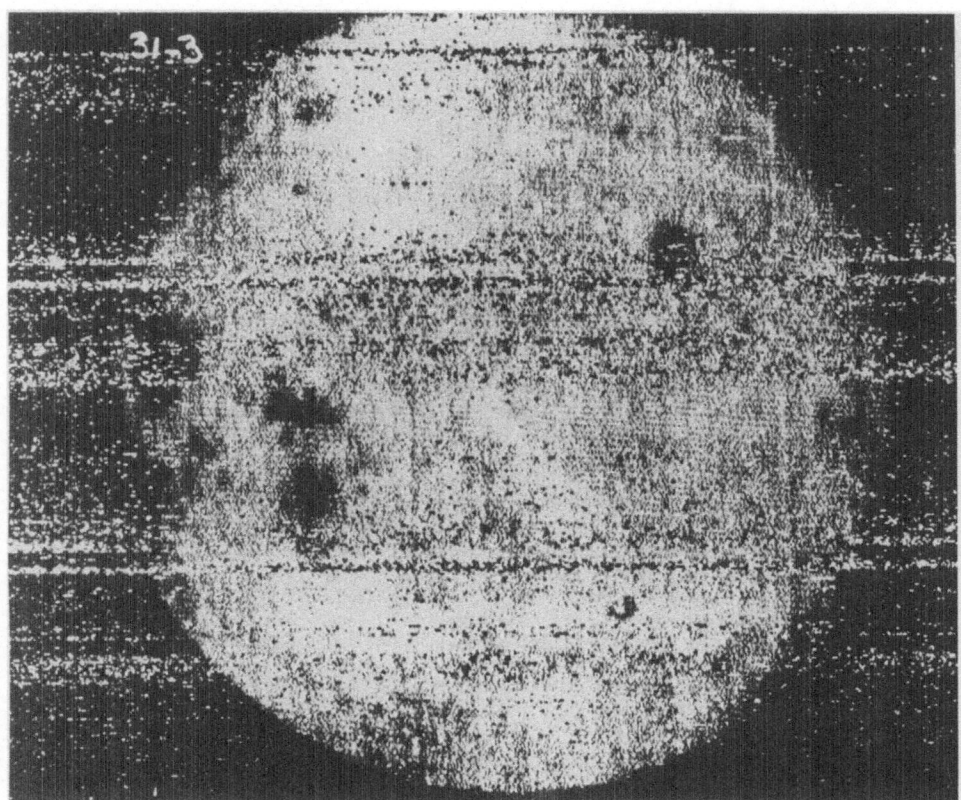

Fig. 10.1.   One of the Luna 3 photographs of the Moon's far side taken by the Soviets in October 1959.
U.S.S.R. Photo

side lunar map were performed in Moscow jointly by the Sternberg Astronomical Institute under the direction of Yu. N. Lipsky and the Central Scientific Research Institute of Geodesy, Aerial Surveying and Cartography under the direction of N. A. Sokolova.

1. Far Side Map

The first map of the far side of the Moon was published in the *Atlas of the Far side of the Moon* edited by N. P. Barabashov, A. A. Mikhailov and Yu. N. Lipsky, March

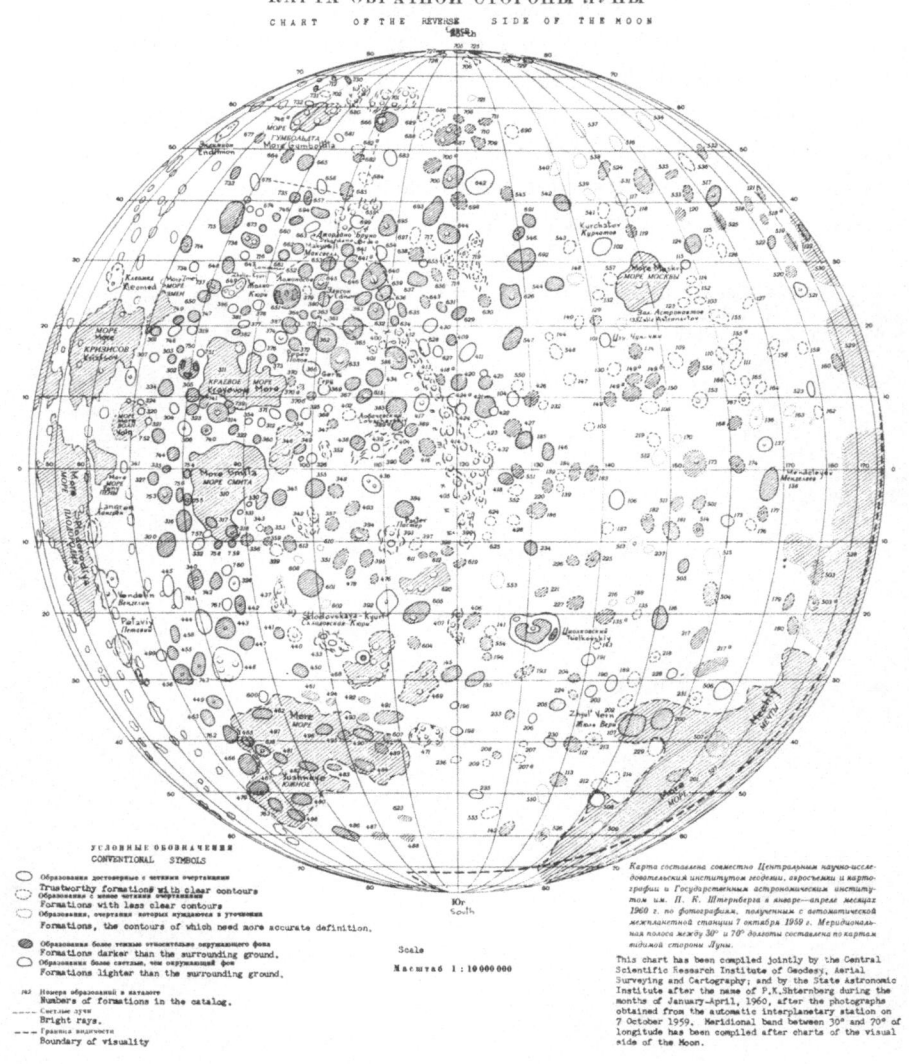

Fig. 10.2.   The first map of the far side of the Moon. It was produced from Luna 3 photographs by U.S.S.R. in March 1960.

1960. This map, shown in Figure 10.2 was compiled on an equatorial orthographic projection with the prime meridian at 120° E. The original, at a scale of 1:10 000 000 at the equator, contains a coordinate latitude-longitude grid at 10° intervals. The meridional band between 30° and 70° E. was based on Wilkins' map which is somewhat unreliable near the limb.

The legend in the lower left margin of Figure 10.2 reads that the solid-outline areas are trustworthy formations with contours (outlines), the dashed features are formations with less clear contours and the dotted outlines are formations, the contours of which need more accurate definition. The shaded features are formations darker than the surrounding area. Leading from the center of the map upward to the left is an enlongated dashed formation identified as the Soviet Mountains (Montes

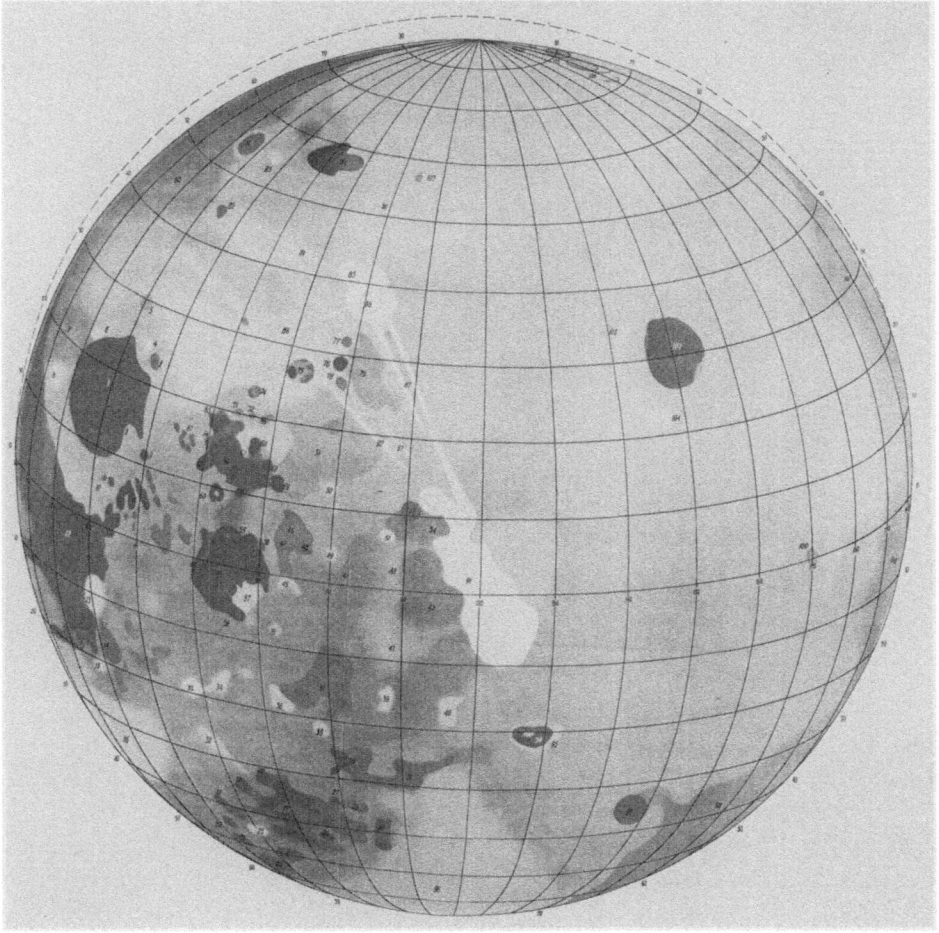

Fig. 10.3.   A schematic chart of the far side of the Moon. It was compiled from Luna 3 photographs by the Pulkovo Main Astronomical Observatory, Leningrad.

Sovietici), subsequently disproven by the Orbiter photographs. The features numbered on this map are identified and described in a listing in the *Atlas of the Far Side of the Moon*. However, the inclusion of too much doubtful detail lessens the value of this map. Eighteen new farside names proposed by the Soviets in this Atlas were adopted by the IAU in 1961 (Berkeley). Among these names were Tsiolkovsky, the Russian father of rocketry, Jules Verne, Pasteur, Edison and Mare Moscoviense. The name 'Montes Sovietici' was dropped in 1970.

From a duplicate set of Luna 3 photographs, Breido and Shchegolev of the Pulkovo Main Astronomical Observatory, Leningrad, under the direction of A. V. Markov produced the schematic chart of the far side shown in Figure 10.3. Of the 107 features mapped on this schematic chart, 56 are related to the far side. In some respects, the

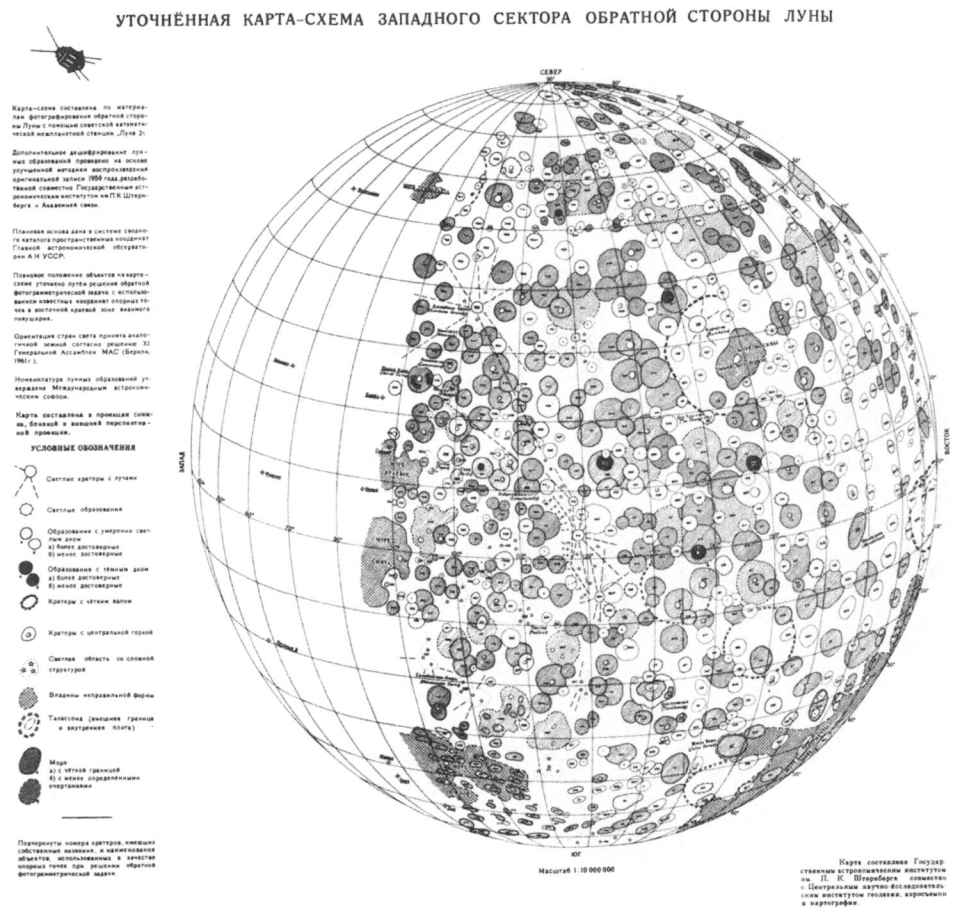

Fig. 10.4. *Improved Sketch Map of the Western Sector of the Far Side of the Moon*. This map was compiled by the P. K. Sternberg Astronomical Institute in collaboration with the Central Scientific Research Institute of Geodesy, Aerial Surveying and Cartography.

schematic map compiled at Pulkovo is superior to the Sternberg map in that an attempt was made to reproduce both the shapes and brightness of the areas of different reflectivity.

Shown in Figure 10.4 is a later Luna 3 far side map titled *Improved Sketch Map of the Western Sector of the Far Side of the Moon*. The border notes state that additional interpretation of lunar formations was made on the basis of improved methods of reproduction of the 1959 original records, worked out jointly by the P. K. Sternberg State Astronomical Institute and the Academy of Communications. This sketch map, 1:10000000 at the equator, is based on the system of reference catalog of space coordinates of the Main Astronomical Observatory of the Ukrainian Academy of Sciences.

Fig. 10.5.   *Photomap of the Visible Hemisphere of the Moon*, published by U.S.S.R. in 1967.

2. Photomap

A photomap of the near side of the Moon, *Photomap of the Visible Hemisphere of the Moon*, shown in Figure 10.5 was published in 1967 at the Sternberg Institute under the direction of Lipsky. It was compiled on the basis of original photographs received from native and foreign observatories, with additional materials from photographic atlases of the Moon.

At a scale of 1 : 5 000 000, this photomap was drawn on an oblique, positive, exterior

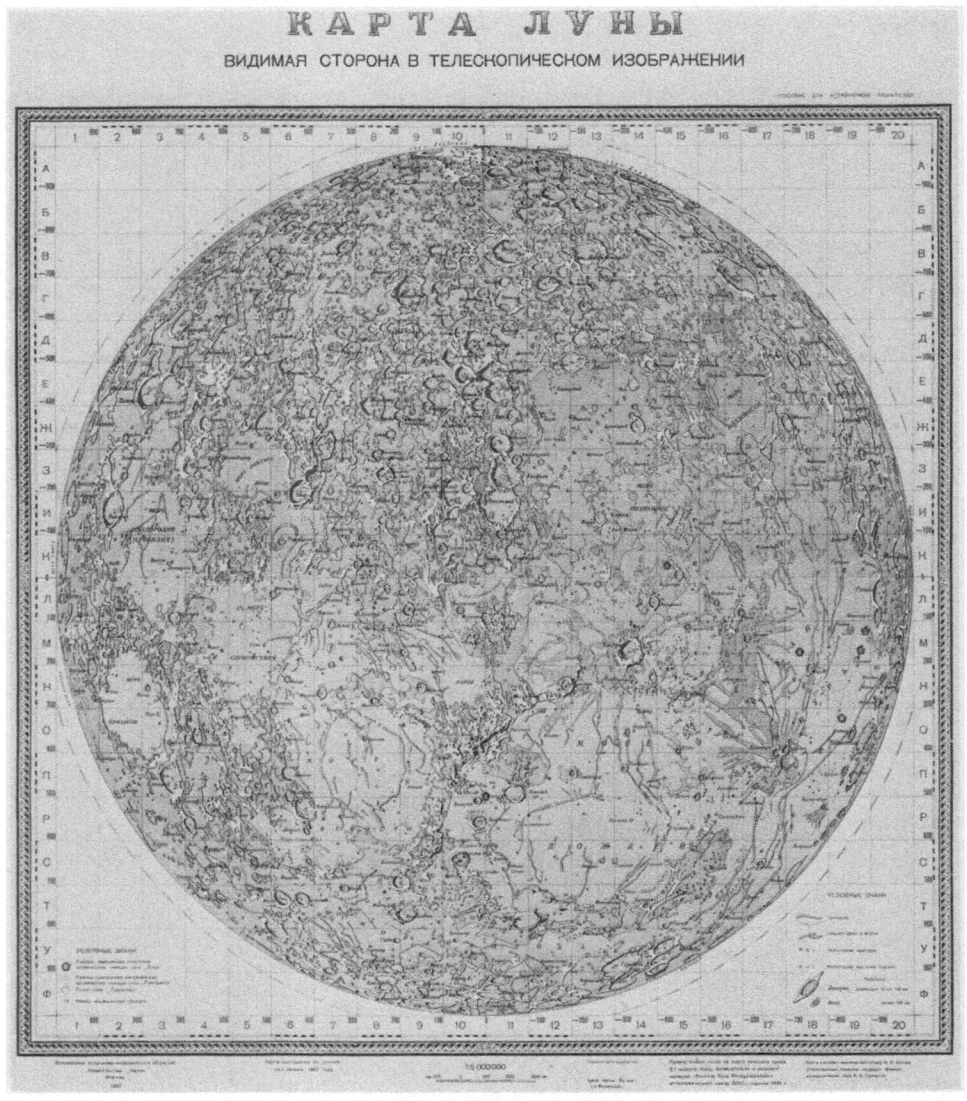

Fig. 10.6.   *Map of the Moon – Telescopic Image of the Near Side*, published by U.S.S.R. in 1967.

perspective projection with positive libration values in latitude and longitude close to the maximum.

Orientation of the cardinal points were taken analogous to the Earth in accordance with resolutions of the XI General Assembly of the IAU (Berkeley, 1961). The photo-map was compiled by mosaic assembly of several photographs. For the eastern portion of the chart, photographs taken in the vicinity of the first quadrant were used. Likewise, the western portion was compiled from last quadrant photographs. The photographs were retouched where the separate sections join.

As the mathematical basis for the grid of selenographic coordinates, reference points were used from the *Summary Catalogue of Spatial Coordinates of 160 Base Points*, by the Main Astronomical Observatory, Academy of Sciences, Ukrainian S.S.R. (1965).

3. Map of the Moon

Shown in Figure 10.6 is a *Map of the Moon*, with a subtitle of *Telescopic Image of the Near Side*. The note in the margin states that it is an Aid for Amateur Astronomers and sells for 62 kopecks with brochure.

This map, 1 : 5 000 000 at the equator, was compiled on an orthographic projection. The rectangular map grid coordinates are plotted at intervals of 0.1 lunar radius, in accordance with the catalog of lunar feature names of the International Astronomical Union, 1935, edition. Shown on the map and identified in the legend are the landing areas of the Soviet Luna, American Ranger and Surveyor spacecraft as of January 1, 1967.

The *Map of the Moon* was compiled by the engineer-cartographer I. I. Katiayev and edited by V. A. Bronshten. It was published in 1967 in Moscow by the National Astronomic-Geodetic Society.

4. 1:1 000 000 Scale Mapping

In 1966 and 1967, the Soviets compiled a series of seven 1:1 000 000 scale maps on the Mercator projection, covering the equatorial zone of the visible hemisphere of the Moon (see Figure 10.7 for index coverage). This series encompasses 8° N–S lat. and

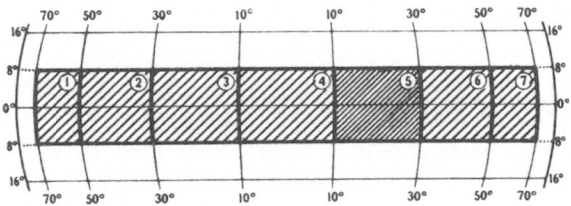

Fig. 10.7.   An index to the U.S.S.R. 1 : 1 000 000 scale Map of the Moon series. Seven maps were published in 1968. Map 5 is shown in Figure 10.8.

70° E–W long. Individual sheets cover 16° of lunar latitude and 20° of longitude. They are numbered from 1 to 7 and named for the region or major feature contained.

Shown in Figure 10.8 is Delambre, Sheet 5 of the 1:1 000 000 scale series. It was published in 1968, based on work performed by the P. K. Sternberg State Astronomical Institute of the M. V. Lomonosov Moscow State University and the Central Scientific Research Institute of Geodesy, Aerial Survey and Cartography.

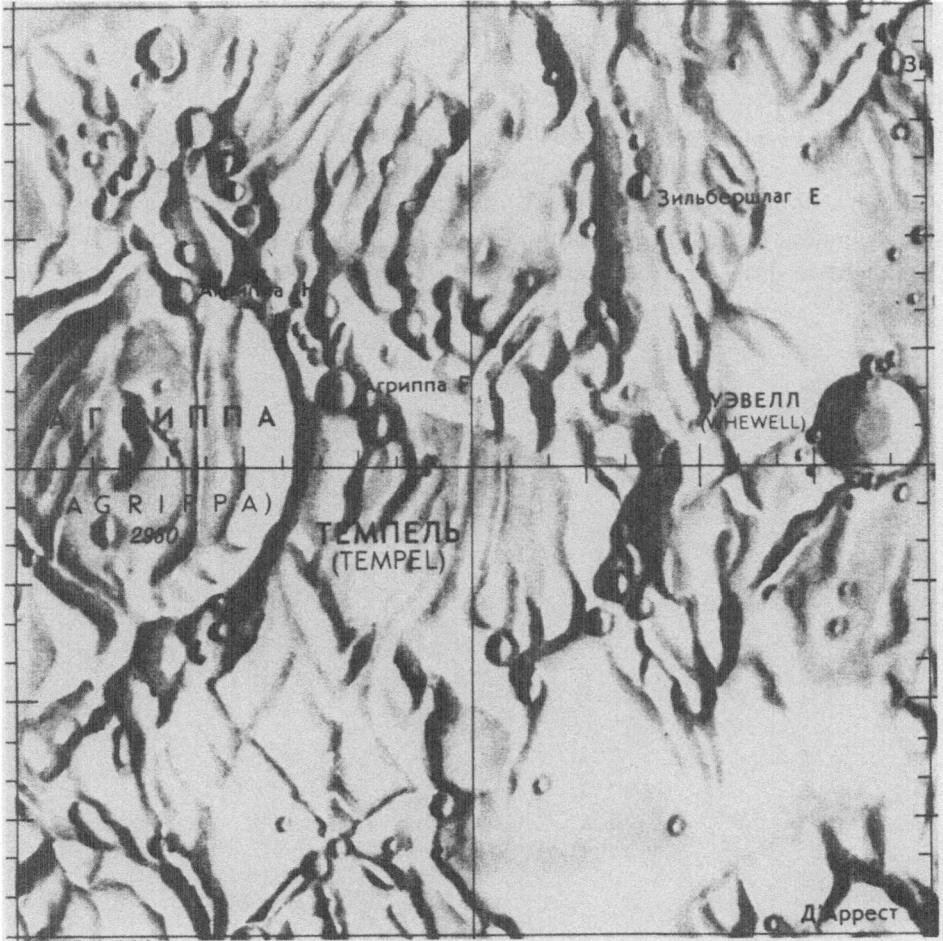

Fig. 10.9.   This is a section of Sheet 5 of the 1:1 000 000 scale Map of the Moon series. Shown at publishing scale are the craters Agrippa, Temple and Whewell.

This series is similar in format, size and content to the NASA Lunar Astronautical Chart (LAC) series. Topography is portrayed by shadient relief. Spot elevations are used to define absolute elevations and heights of prominent features above or below surrounding terrains. Marginal notes state that basis for its horizontal and vertical control is a catalog of space coordinates produced by the Main Astronomical Observatory of the Academy of Sciences, Ukrainian S.S.R. Relative heights are derived from measurements of Soviet and foreign observatories.

Nomenclature of lunar features, including Latin translations corresponds to resolutions of the General Assemblies of the International Astronomical Union V (Paris, 1935), XI (Berkeley, 1961) and XII (Hamburg, 1964).

Figure 10.9 shows a section of Sheet 5 at reproduction scale covering the craters Agrippa, Temple and Whewell. Features are printed in brown.

5. 1:5000000 Scale Mapping

In 1967, Sternberg and the Topogeodetic Service of U.S.S.R. published a complete lunar map in nine sheets which sells for 2 rubles, 18 kopecks. Six sheets provide latitudinal coverage from 60° N to 60° S on a 1:5000000 scale modified cylindrical projection. Another sheet contains the north and south polar areas on a 1:10000000 scale azimuthal projection with latitude coverage from the poles to 50° N and 50° S. The other two sheets provide a listing of feature names in Russian and Latin spelling

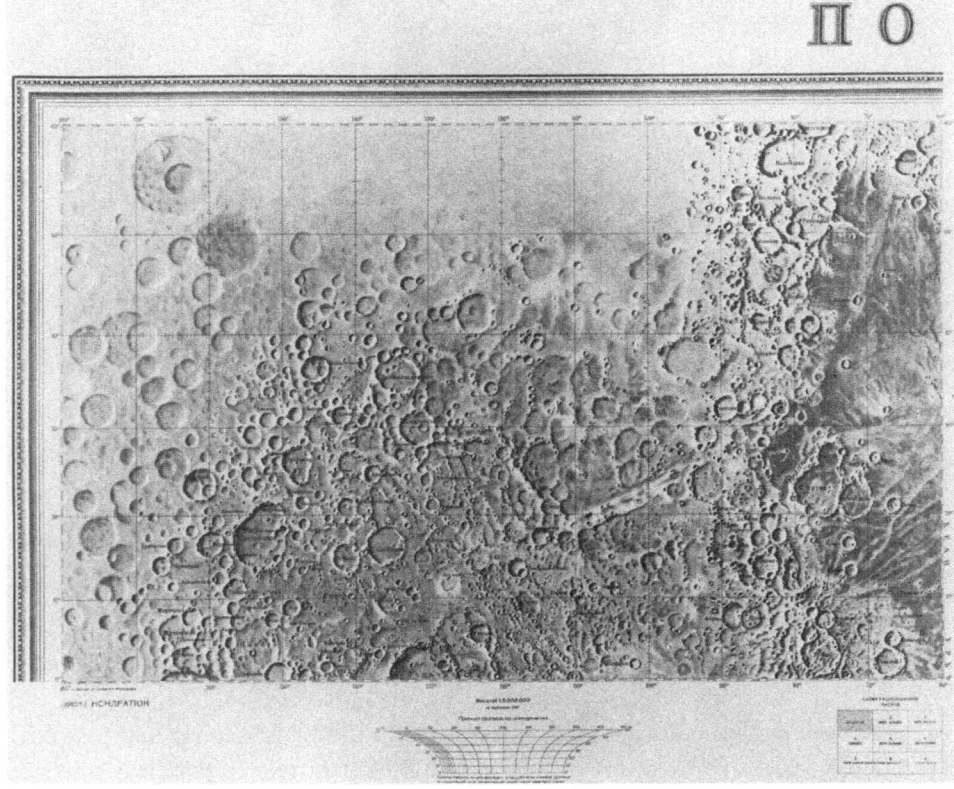

Fig. 10.10.   Sheet 1 of the 1:5000000 scale *Complete Chart of the Moon*, published by U.S.S.R. in 1967. This sheet covers the area from 0° to 60° N lat. and from 60° W long. to the 180th meridian.

for each map sheet. All nine sheets have a common border which permits joining and mounting as a single wall map. The title, in large letters spread across the top three sheets reads, *Complete Chart of the Moon.*

Shown in Figures 10.10 and 10.11 are Sheets 1 and 4 with coverage from 60° W long. to the 180th° meridian. On these two sheets the topographic detail west of 90° longitude was compiled from Zond 3 photographs taken by U.S.S.R. on July 20, 1965.

Fig. 10.11.   Sheet 4 of the 1:5 000 000 scale *Complete Chart of the Moon*, published by U.S.S.R. in 1967. This sheet covers the area from 0° to 60° S lat. and from 60° W long. to the 180th meridian.

One of the 25 Zond 3 images is shown in Figure 10-12. In the newly compiled area on this Soviet map, Lipsky added approx 225 new names which were not IAU approved. These names were subsequently considered by the IAU in 1970. Some were approved, some were relocated to better defined features and the remainder were dropped.

Sheets 2 and 5 which cover the near side from 60° E to 60° W long. were compiled from earth-based photographs and are fairly accurate in both position and feature portrayal. The remaining two sheets, 3 and 6, in the area of Luna 3 photo coverage, are very sketchy and lack reliable feature detail as will be noted in Sheet 3 in Figure 10.13.

In 1969, a revised edition of the 1:5 000 000 nine-sheet series was issued with 99.5% coverage of the lunar surface. The source used to compile the remaining far side coverage was the Orbiter photographs as noted in the map margin. Figure 10.14 shows

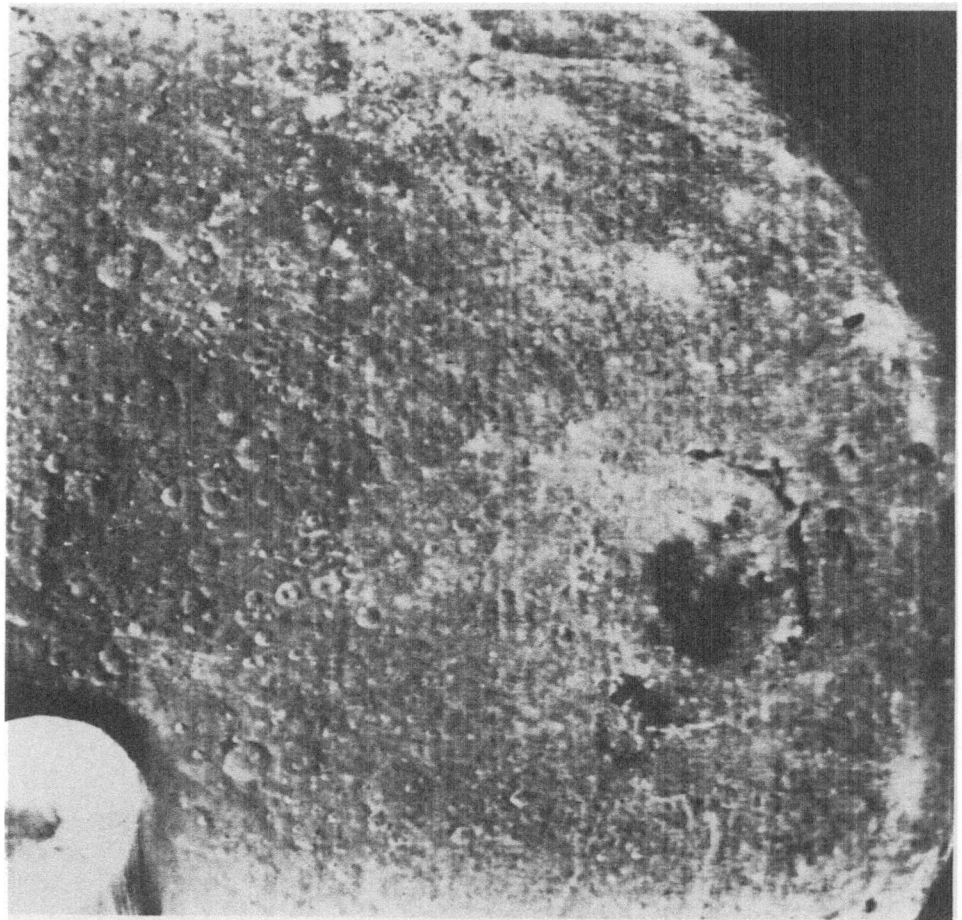

Fig. 10.12.   One of the Zond 3 photographs taken by U.S.S.R. on July 20, 1965. The small dark flat area
to the right of center is the floor of Mare Orientale.   *U.S.S.R. Photo*

the revised Sheet 3 which can be compared to a lack of detail in the earlier edition
shown in Figure 10.13.

Also in the 1969 edition, a change was made in the two polar maps. Both were
enlarged to the scale of 1:5 000 000 and reduced in area to 60° N–S lat. Shown in
Figure 10.15 is the south polar map of this later edition drawn on an azimuthal
projection.

The topography on both editions is drawn by shaded relief and printed in brown
which seems to be the standard color for publishing all U.S.S.R. lunar maps. However,
the 1969 edition carries a light blue tint under the maria.

6. U.S.S.R. Lunar Globe

The Soviets produced their first globe of the Moon in 1961 based on telescopic photographs for the near side and Luna 3 photographs for the far side. However, the far side area between 90° W and 160° W longitude remained blank since the features in this area were unknown at that time.

In 1967, the U.S.S.R. issued another lunar globe which included the far side area photographed by Zond 3. This globe, shown in Figure 10.16 is at a scale of 1:10 000 000 (14 in. in diam). It was published by the Sternberg State Astronomical Institute and the Topographic and Geodetic Service. The compilation was supervised by Yu. N.

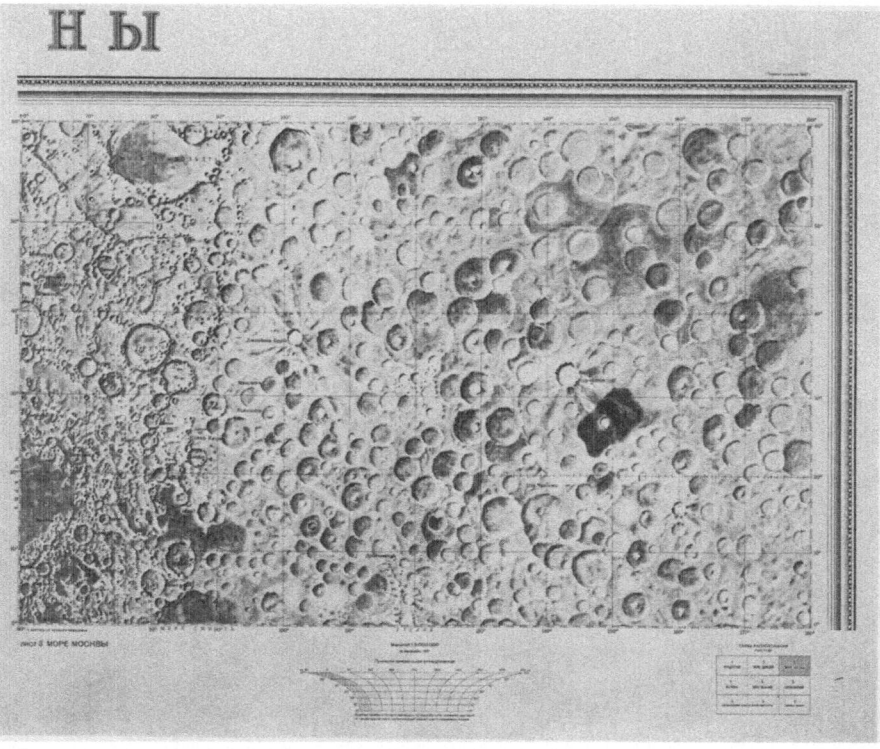

Fig. 10.13. Sheet 3 of the 1:5 000 000 scale *Complete Chart of the Moon*, published by U.S.S.R. in 1967. This sheet covers the area from 0° to 60° N lat. and from 60° E long. to the 180th meridian. The detail is very sketchy in the area of Luna 3 photo coverage.

Lipsky. Assembly of the globe was made at the Moscow Nature and School Works.

The 1967 Russian lunar globe is mounted on a pentagonal black plastic stand. It rests free on a supporting shaft which permits rotation and reversal of the poles. Surface features are portrayed in brown. A latitude/longitude projection at 10° intervals and the names of a majority of the primary features (in English) are over-printed in black. On the far side in the area of Zond 3 coverage, Lipsky included the same unofficial names added to the 1967 U.S.S.R. 1:10 000 000 scale lunar map.

A note printed on the U.S.S.R. lunar globe reads that it was made from the photographs obtained by the Automatic Interplanetary Station Luna 3 on October 7, 1959

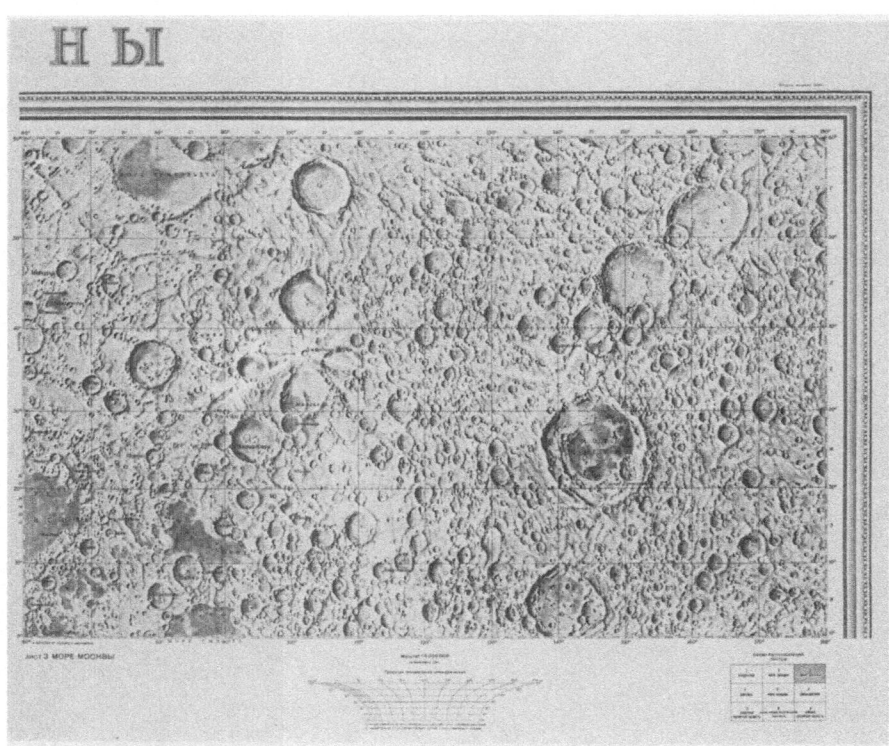

Fig. 10.14.   1969 edition of Sheet 3 of the 1:5 000 000 scale *Complete Chart of the Moon*. Note increased detail from the 1967 edition shown in Figure 10.13.

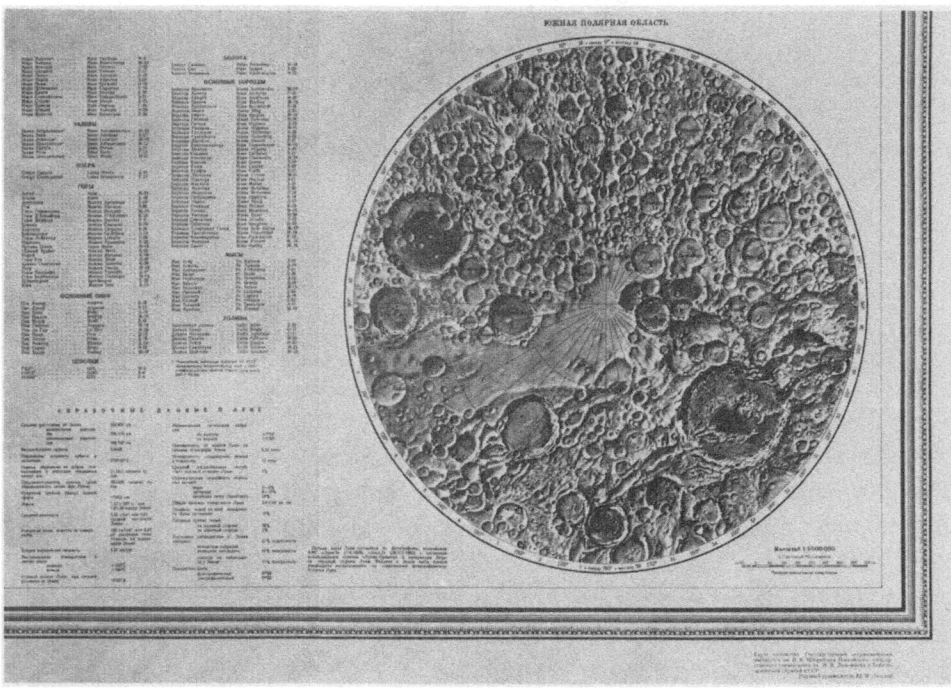

Fig. 10.15.   South pole area of the 1:5 000 000 scale *Complete Chart of the Moon*, published by U.S.S.R. in 1969.

Fig. 10.16.   The 14-in. U.S.S.R. Lunar Globe published in 1967. This globe was based on Luna 3 and Zond 3 photographs. In 1969 it was reissued with increased detail obtained from Orbiter photographs.

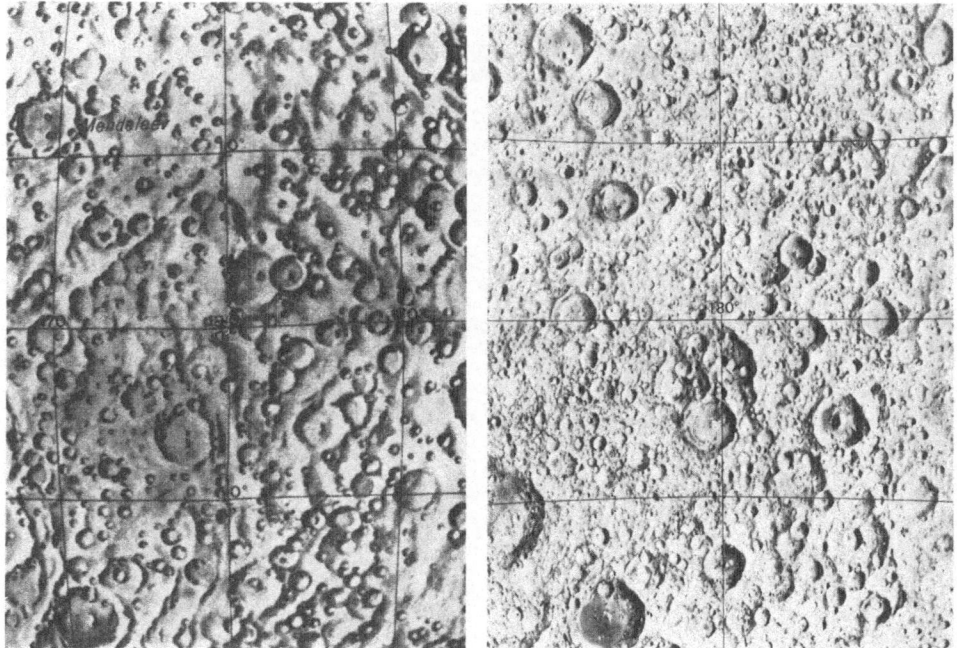

Fig. 10.17.   Comparison of detail shown on the U.S.S.R. Lunar Globe (left) with the NASA Lunar
Globe (right). These two areas are centered on 0° lat. and 180° long.

and Zond 3 on July 20, 1965. The visible side of the Moon was reproduced from
recent photographic lunar atlases.

   In 1969, the Soviets reissued their 1967 lunar globe to include detail compiled from
the NASA Orbiter photographs. The drawing of features on all three globes is basically schematic and does not present many small distinguishing characteristics of
features. This condition is more prevalent on the far side as noted in Figure 10.17
which compares a like area on the U.S.S.R. globe with the NASA globe; however,
some allowance must be made due to differences in scale between the two globes.

Reference

Barabashov, N. P., Mikhailov, A. A., and Lipsky, Yu. N.: 1960, *Atlas of the Far Side of the Moon*, U.S.S.R.
   Academy of Sciences, Moscow.

CHAPTER 11

NATIONAL GEOGRAPHIC LUNAR MAPPING

One of the best and most popular lunar maps of the space age was produced in 1969 by the National Geographic Society (NGS), Washington, D.C. Measuring 28″ × 42″ it was designed by National Geographic staff cartographer, David Cook. He chose to use the Lambert azimuthal equal-area projection, which shows each feature in its true direction, or azimuth, from the center of its hemisphere, and in true area scale. Titled *The Earth's Moon* it is printed in a silvery-gray against the blue-blackness of night (Figure 11.1).

Contained within its borders is a wealth of basic information about the Moon, such as Physical Features, Physical Properties, Tides, Centrifugal Force, Phases of the Moon, Moon Lore, The Lunar Month, Libration, Total Solar, Annual Solar and Lunar Eclipses.

Shown across the bottom are eleven drawings of the Moon phases beginning with the waxing crescent, followed by the first quarter, waxing gibbous, full Moon, waning gibbous, last quarter, and finally, the waning crescent. These drawings were made from telescopic photographs of the Moon taken at these phases. Stretched across the full width at the bottom of the map is a flight plan of Apollo from Earth to the Moon and return. This is drawn to a true relative scale.

Shown in red are the final resting places of all but one of the 23 unmanned spacecraft that had reached the Moon's surface by 1969 – 17 American and 6 Soviet. The crash site of U.S. Orbiter 4 is not certain.

The names of the prominent features of the Moon have been shown and indexed in columns on the right and left borders. The names can be located by a coordinate scheme of red letters on the central meridian of each hemisphere and red numbers across the lunar equator. Lacking are the official names on the far side which were not approved by the International Astronomical Union until 1970.

To give its map an intriguing border, the National Geographic Society assembled 163 names and printed them end to end. This list includes astronauts, cosmonauts, spacecraft, scientists, lunar feature names, and terms related to the Moon, Moon lore and space travel. Thus, it is interesting to note names such as Alan Sheppard, Diana, Orbiter 5, Luna 13, Perilune, Mary Blagg, Julius Schmidt, Cape Kennedy, Ocean of Storms, Yuri Gagarin, etc., as they tell the courageous story of man's conquest of space.

U.S. Lunar Orbiter photographs taken in 1966 and 1967 were used as the prime source for both control and surface features. One of the biggest problems faced by NGS was a lack of control for the near side limb regions and for the far side. This

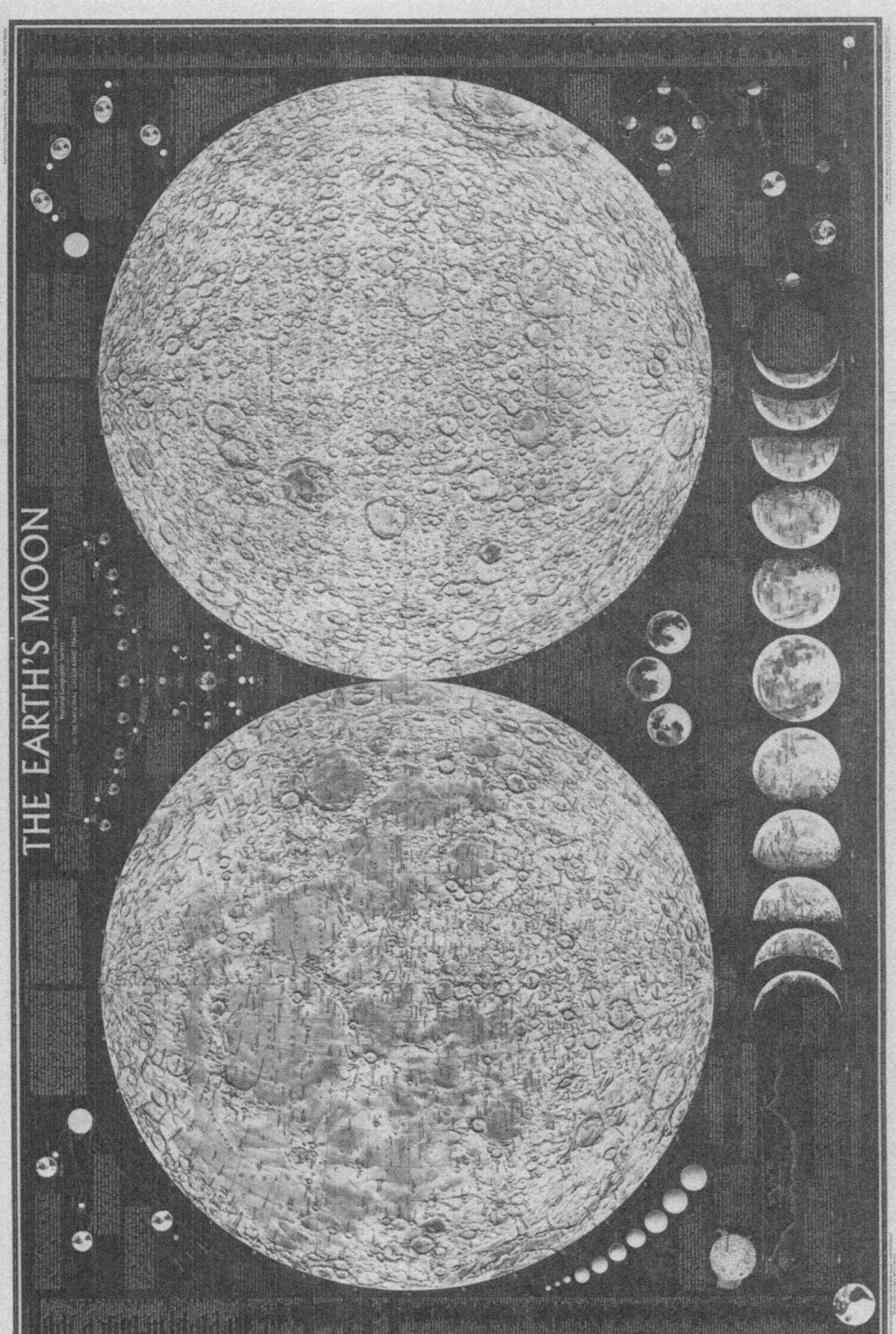

Fig. 11.1. *The Earth's Moon*, a unique lunar wall map, published by the National Geographic Society in 1969.

Fig. 11.2.   Using a 40-in. Earth globe to represent the Moon, National Geographic staff cartographer David Cook checks the latitude-longitude position before photographing it from the same relative altitude and position from which the Orbiter picture was taken.   *NGS Photo*

Fig. 11.3. The National Geographic Society built this ingenious rig to determine the exact position of features on their lunar map. David Cook is holding a slide rule, Richard Furno is aligning the Earth globe while Victor Boswell, Jr. positions the light. The picture thus obtained establishes the precise latitudes and longitudes of lunar features shown in the Orbit view. *NGS Photo*

problem was solved by building an ingenious rig holding a 40-in. Earth globe containing lines of latitude and longitude to represent the Moon (see Figure 11.2). Its object was to photograph a perspective view of the Earth globe in the same position as each orbit photograph was taken from space. Then, with the camera shown in the

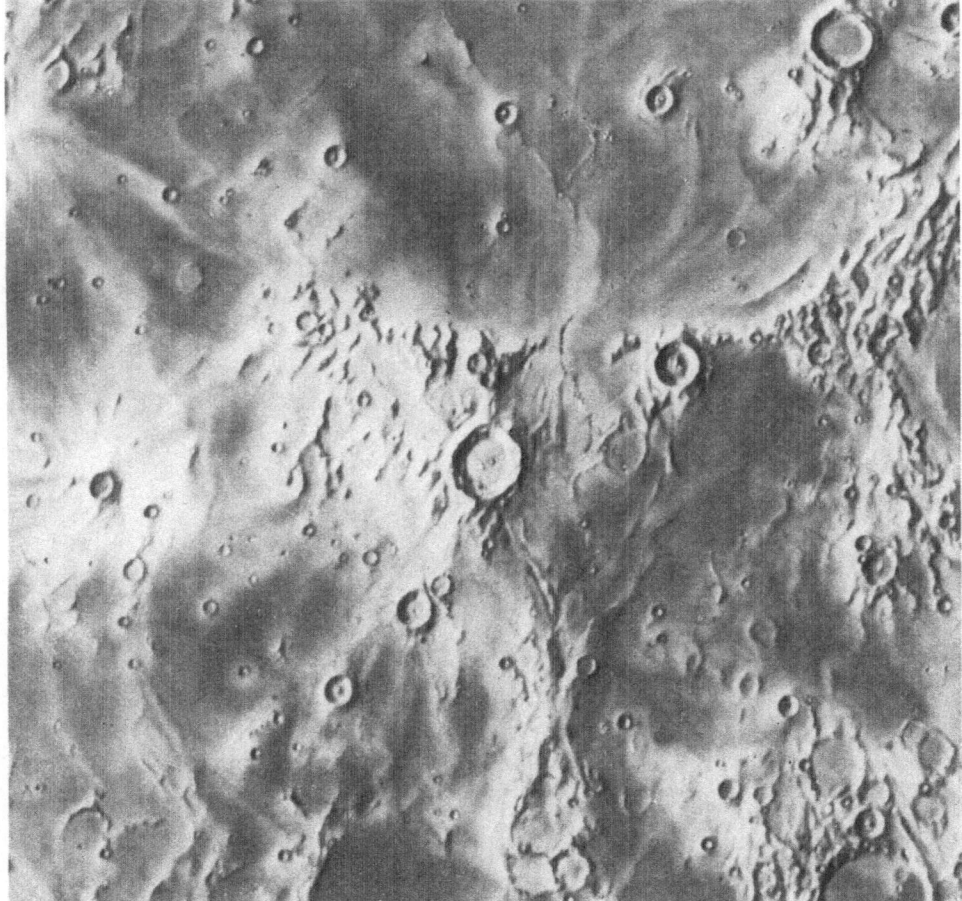

Fig. 11.4. This is a section of the National Geographic lunar map shown at reproduction scale. Note the subtle feature detail portrayed in this drawing.

foreground of Figure 11.3, NGS took hundreds of pictures, each from a different and calculated point in relation to the globe. The resulting perspective grids when superimposed on the orbit photographs allowed their cartographers to plot the selenographic position of each feature.

Shown in Figure 11.4 is a section of the near side drawing centered on the crater

Fig. 11.5.   National Geographic staff artist Tibor Toth uses an airbrush to capture as much of the lunar
surface as the scale of the map would permit.   *NGS Photo*

Copernicus. This drawing, at the same scale as the published map, shows all features
as if lighted by a setting Sun. The portrayal of lunar features bears a close resemblance
to the technique developed by ACIC. This was no coincidence since their artist, Tibor
Toth (Figure 11.5) spent several weeks in training at Lowell Observatory where the
ACIC lunar charts were drawn.

Reference

Cook, David W.: 1969, 'How We Mapped the Moon', *National Geographic*, February.

NAME INDEX

ASTROPHYSICS AND SPACE SCIENCE LIBRARY

Edited by

J. E. Blamont, R. L. F. Boyd, L. Goldberg, C. de Jager, Z. Kopal, G. H. Ludwig, R. Lüst,
B. M. McCormac, H. E. Newell, L. I. Sedov, Z. Švestka, and W. de Graaff

1. C. de Jager (ed.), *The Solar Spectrum. Proceedings of the Symposium held at the University of Utrecht, 26–31 August, 1963.* 1965, XIV + 417 pp.
2. J. Ortner and H. Maseland (eds.), *Introduction to Solar Terrestrial Relations. Proceedings of the Summer School in Space Physics held in Alpbach, Austria, July 15–August 10, 1963 and Organized by the European Preparatory Commission for Space Research.* 1965, IX + 506 pp.
3. C. C. Chang and S. S. Huang (eds.), *Proceedings of the Plasma Space Science Symposium, held at the Catholic University of America, Washington, D.C., June 11–14, 1963.* 1965, IX + 377 pp.
4. Zdeněk Kopal, *An Introduction to the Study of the Moon.* 1966, XII + 464 pp.
5. B. M. McCormac (ed.), *Radiation Trapped in the Earth's Magnetic Field. Proceedings of the Advanced Study Institute, held at the Chr. Michelsen Institute, Bergen, Norway, August 16–September 3, 1965.* 1966, XII + 901 pp.
6. A. B. Underhill, *The Early Type Stars.* 1966, XII + 282 pp.
7. Jean Kovalevsky, *Introduction to Celestial Mechanics,* 1967, VIII + 427 pp.
8. Zdeněk Kopal and Constantine L. Goudas (eds.), *Measure of the Moon. Proceedings of the Second International Conference on Selenodesy and Lunar Topography, held in the University of Manchester, England, May 30–June 4, 1966.* 1967, XVIII + 479 pp.
9. J. G. Emming (ed.), *Electromagnetic Radiation in Space. Proceedings of the Third ESRO Summer School in Space Physics, held in Alpbach, Austria, from 19 July to 13 August, 1965.* 1968, VIII + 307 pp.
10. R. L. Carovillano, John F. McClay, and Henry R. Radoski (eds.), *Physics of the Magnetosphere, Based upon the Proceedings of the Conference held at Boston College, June 19–28, 1967.* 1968, X + 686 pp.
11. Syun-Ichi Akasofu, *Polar and Magnetospheric Substorms.* 1968, XVIII + 280 pp.
12. Peter M. Millman (ed.), *Meteorite Research. Proceedings of a Symposium on Meteorite Research, held in Vienna, Austria, 7–13 August, 1968.* 1969, XV + 941 pp.
13. Margherita Hack (ed.), *Mass Loss from Stars. Proceedings of the Second Trieste Colloquium on Astrophysics, 12–17 September, 1968.* 1969, XII + 345 pp.
14. N. D'Angelo (ed.), *Low-Frequency Waves and Irregularities in the Ionosphere. Proceedings of the 2nd ESRIN-ESLAB Symposium, held in Frascati, Italy, 23–27 September, 1968.* 1969, VII + 218 pp.
15. G. A. Partel (ed.), *Space Engineering. Proceedings of the Second International Conference on Space Engineering, held at the Fondazione Giorgio Cini, Isola di San Giorgio, Venice, Italy, May 7–10, 1969.* 1970, XI + 728 pp.
16. S. Fred Singer (ed.), *Manned Laboratories in Space. Second International Orbital Laboratory Symposium.* 1969, XIII + 133 pp.
17. B. M. McCormac (ed.), *Particles and Fields in the Magnetosphere. Symposium Organized by the Summer Advanced Study Institute, held at the University of California, Santa Barbara, Calif., August 4–15, 1969.* 1970, XI + 450 pp.
18. Jean-Claude Pecker, *Experimental Astronomy.* 1970, X + 105 pp.
19. V. Manno and D. E. Page (eds.), *Intercorrelated Satellite Observations related to Solar Events. Proceedings of the Third ESLAB/ESRIN Symposium held in Noordwijk, The Netherlands, September 16–19, 1969.* 1970, XVI + 627 pp.
20. L. Mansinha, D. E. Smylie, and A. E. Beck, *Earthquake Displacement Fields and the Rotation of the Earth. A NATO Advances Study Institute Conference Organized by the Department of Geophysics, University of Western Ontario, London, Canada, June 22–28, 1969.* 1970, XI + 308 pp.
21. Jean-Claude Pecker, *Space Observatories.* 1970, XI + 120 pp.
22. L. N. Mavridis (ed.), *Structure and Evolution of the Galaxy. Proceedings of the NATO Advanced Study Institute, held in Athens, September 8–19, 1969.* 1971, VII + 312 pp.
23. A. Muller (ed.), *The Magellanic Clouds. A European Southern Observatory Presentation: Principal*

Prospects, Current Observational and Theoretical Approaches, and Prospects for Future Research. Based on the Symposium on the Magellanic Clouds, held in Santiago de Chile, March 1969, on the Occasion of the Dedication of the European Southern Observatory. 1971, XII + 189 pp.

24. B. M. McCormac (ed.), *The Radiating Atmosphere. Proceedings of a Symposium Organized by the Summer Advanced Study Institute, held at Queen's University, Kingston, Ontario, August 3–14, 1970.* 1971, XI + 455 pp.

25. G. Fiocco (ed.), *Mesopheric Models and Related Experiments. Proceedings of the 4th ESRIN–ESLAB Symposium, held at Frascati, Italy, July 6–10, 1970.* 1971, VIII + 298 pp.

26. I. Atanasijević, *Selected Exercises in Galactic Astronomy.* 1971, XII + 144 pp.

27. C. J. Macris (ed.), *Physics of the Solar Corona. Proceedings of the NATO Advanced Study Institute on Physics of the Solar Corona, held at Cavouri-Vouliagmeni, Athens, Greece, 6–17 September 1970.* 1971, XII + 345 pp.

28. F. Delobeau, *The Environment of the Earth.* 1971, IX + 113 pp.

29. E. R. Dyer (general ed.), *Solar-Terrestrial Physics 1970. Proceedings of the International Symposium on Solar-Terrestrial Physics, held in Leningrad, U.S.S.R., 12–19 May 1970.* 1972, VIII + 938 pp.

30. V. Manno and J. Ring (eds.), *Infrared Detection Techniques for Space Research, Proceedings of the Fifth ESLAB-ESRIN Symposium, held in Noordwijk, The Netherlands, June 8–11, 1971.* 1972, XII + 344 pp.

31. M. Lecar (ed.), *Gravitational N-Body Problem, Proceedings of IAU Colloquium No. 10, held in Cambridge, England, August 12–15, 1970.* 1972, XI + 441 pp.

32. B. M. McCormac (ed.), *Earth's Magnetospheric Processes. Proceedings of a Symposium Organized by the Summer Advanced Study Institute and Ninth ESRO Summer School, held in Cortina, Italy, August 30–September 10, 1971.* 1972, VIII + 417 pp.

33. Antonin Rükl, *Maps of Lunar Hemispheres.* 1972, V + 24 pp.

34. V. Kourganoff, *Introduction to the Physics of Stellar Interiors.* 1973, XI + 115 pp.

35. B. M. McCormac (ed.), *Physics and Chemistry of Upper Atmospheres. Proceedings of Symposium Organized by the Summer Advanced Study Institute, held at the University of Orléans, France, July 31–August 11, 1972.* 1973, VIII + 389 pp.

36. J. D. Fernie (ed.), *Variable Stars in Globular Clusters and in Related Systems. Proceedings of the IAU Colloquim No. 21, held at the University of Toronto, Toronto, Canada, August 29–31, 1972.* 1973, IX + 234 pp.

37. R. J. L. Grard (ed.), *Photon and Particle Interaction with Surfaces in Space. Proceedings of the 6th ESLAB Symposium, held at Noordwijk, the Netherlands, 26–29 September, 1972,* 1973, XV + 577 pp.

38. Werner Israel (ed.), *Relativity, Astrophysics and Cosmology. Proceedings of the Summer School, held 14–26 August, 1972, at the BANFF Centre, BANFF, Alberta, Canada.* 1973. IX + 323 pp.

39. B. D. Tapley and V. Szebehely (eds.), *Recent Advances in Dynamical Astronomy. Proceedings of the NATO Advanced Study Institute in Dynamical Astronomy, held in Cortina d'Ampezzo, Italy, August 9–12, 1972.* 1973, XIII + 468 pp.

40. A. G. W. Cameron (ed.), *Cosmochemistry. Proceedings of the Symposium on Cosmochemistry, held at the Smithsonian Astrophysical Observatory, Cambridge, Mass., August 14–16, 1972.* 1973, X + 173 pp.

41. M. Golay, *Introduction to Astronomical Photometry.* 1974, approx. 570 pp.

42. D. E. Page (ed.), *Correlated Interplanetary and Magnetospheric Observations. Proceedings of the Seventh ESLAB Symposium, held at Saulgau, W. Germany, 22–25 May, 1973.* 1974, XIV + 662 pp.

45. C. B. Cosmovici (ed.), *Supernovae and Supernova Remnants. Proceedings of the International Conference on Supernovae, held in Lecce, Italy, May 7–11, 1973.* 1974, XVII + 387 pp.

In preparation:

43. R. Giacconi and H. Gursky (eds.), *X-Ray Astronomy.*

44. B. M. McCormac (ed.), *Magnetospheric Physics. Proceedings of the Advanced Summer Institute, held at Sheffield, U.K., August 1973.*

46. A. P. Mitra, *Ionospheric Effects of Solar Flares.*